Beast ACADEMY

By Art of Problem Solving

MATH
PRACTICE
3D

Published by: AoPS Incorporated
                  10865 Rancho Bernardo Rd Ste 100
                  San Diego, CA 92127-2102
                  info@BeastAcademy.com

ISBN: 978-1-934124-47-5

Written by Jason Batterson and Shannon Rogers
Book Design by Lisa T. Phan
Illustrations by Erich Owen
Grayscales by Greta Selman

Visit the Beast Academy website at www.BeastAcademy.com.
Visit the Art of Problem Solving website at www.artofproblemsolving.com.
Printed in the United States of America.
2019 Printing.

Contents:

This is Practice Book 3D in the Beast Academy level 3 series.

3A
• Shapes
• Skip-Counting
• Perimeter and Area

3B
• Multiplication
• Perfect Squares
• The Distributive Property

3C
• Variables
• Division
• Measurement

3D
• Fractions
• Estimation
• Area

For more resources and information, visit BeastAcademy.com/resources.

This is Beast Academy Practice Book 3D.

MATH PRACTICE 3D

Each chapter of this Practice book corresponds to a chapter from Beast Academy Guide 3D.

MATH GUIDE 3D

The first page of each chapter includes a recommended sequence for the Guide and Practice book.

You may also read the entire chapter in the Guide before beginning the Practice chapter.

Use this Practice book with Guide 3D from BeastAcademy.com.

Recommended Sequence:

Book	Pages
Guide:	12 – 24
Practice:	7 – 11
Guide:	25 – 33
Practice:	12 – 21
Guide:	34 – 44
Practice:	22 – 38
Guide:	45 – 51
Practice:	39 – 45

You may also read the entire chapter in the Guide before beginning the Practice chapter.

Some problems in this book are very challenging. These problems are marked with a ★. The hardest problems have two stars!

Every problem marked with a ★ has a *hint!*

Hints for the starred problems begin on page 104.

Other problems are marked with a ✏. For these problems, you should write an explanation for your answer.

Some pages direct you to related pages from the Guide.

None of the problems in this book require the use of a calculator.

Solutions are in the back, starting on page 108.

A complete explanation is given for every problem!

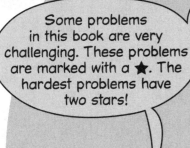

CHAPTER 10
Fractions

Use this Practice book with
Guide 3D from BeastAcademy.com.

Recommended Sequence:

You may also read the entire chapter
in the Guide before beginning the
Practice chapter.

A *fraction* is a number.

Fractions are also another way to write division.

For example, we can write $1 \div 2$ as $\frac{1}{2}$.

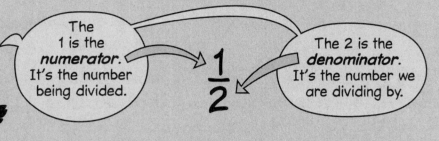

The 1 is the *numerator*. It's the number being divided.

$\frac{1}{2}$

The 2 is the *denominator*. It's the number we are dividing by.

Since a fraction is a number, we can put it on the number line.

To locate $\frac{1}{2}$ on the number line, we divide the number line between 0 and 1 into two pieces of equal length.

Each piece has a length of $\frac{1}{2}$.

The first piece begins at 0 and ends at $\frac{1}{2}$.

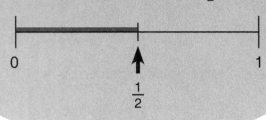

PRACTICE | Label the number marked with an arrow on each number line below.

1.

2.

3.

4.
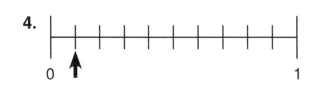

A **unit fraction** is a fraction that has a numerator of 1.

We can compare unit fractions on a number line.

EXAMPLE | Which is greater, $\frac{1}{5}$ or $\frac{1}{7}$?

We can compare $\frac{1}{5}$ to $\frac{1}{7}$ on the number line.

To locate $\frac{1}{5}$, we divide the number line between 0 and 1 into five equal pieces. The first piece begins at 0 and ends at $\frac{1}{5}$.

To locate $\frac{1}{7}$, we divide the number line between 0 and 1 into seven equal pieces. The first piece begins at 0 and ends at $\frac{1}{7}$.

Since $\frac{1}{5}$ is to the right of $\frac{1}{7}$ on the number line, $\frac{1}{5}$ is greater than $\frac{1}{7}$.

The more equal pieces you divide something into, the smaller each piece must be!

PRACTICE | Compare each pair of fractions given below.

5. Which is greater, $\frac{1}{2}$ or $\frac{1}{3}$?

5. _____

6. Which is greater, $\frac{1}{6}$ or $\frac{1}{9}$?

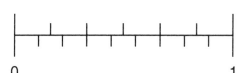

6. _____

7. Which is greater, $\frac{1}{11}$ or $\frac{1}{10}$?

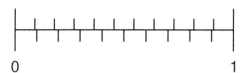

7. _____

8. Which is greater, $\frac{1}{40}$ or $\frac{1}{29}$?

8. _____

9. Which is greater, $\frac{1}{91}$ or $\frac{1}{100}$?

9. _____

Some fractions equal whole numbers!

EXAMPLE | Write $\frac{15}{3}$ as a whole number.

Since fractions are another way to write division, $\frac{15}{3}$ means $15 \div 3$.

So, $\frac{15}{3}$ equals $15 \div 3 = \mathbf{5}$.

PRACTICE | Write each fraction below as a whole number.

10. $\frac{16}{8} =$ _____

11. $\frac{48}{4} =$ _____

12. $\frac{45}{9} =$ _____

13. $\frac{75}{25} =$ _____

14. $\frac{39}{13} =$ _____

15. $\frac{12}{12} =$ _____

PRACTICE | Fill in the numerator that will make each equation below true.

16. $\dfrac{}{3} = 2$

17. $\dfrac{}{7} = 10$

18. $\dfrac{}{3} = 12$

19. $\dfrac{}{6} = 9$

PRACTICE | Fill in the denominator that will make each equation below true.

20. $\dfrac{12}{} = 2$

21. $\dfrac{56}{} = 7$

22. $\dfrac{35}{} = 5$

23. $\dfrac{36}{} = 4$

In these Fraction Link puzzles, we connect fractions and whole numbers that are equal.

In a **Fraction Link** puzzle, the goal is to connect each pair of equal numbers by a path.

- Paths may only go up, down, left or right through squares.
- Paths must begin and end at a number, but they may not pass *through* squares that contain numbers.
- Only one path may pass through each square.

Below is an example of a Fraction Link puzzle and its solution:

PRACTICE | Solve each Fraction Link puzzle below. We recommend using a pencil.

You can print more copies of these Fraction Link puzzles at BeastAcademy.com.

24.

1			
2		$\frac{10}{5}$	
3		$\frac{9}{3}$	
$\frac{6}{6}$			

25.

1			2
	$\frac{6}{2}$	$\frac{5}{5}$	
	$\frac{4}{2}$		
			3

26.

$\frac{8}{2}$			
	2		
	3		
	4		
$\frac{6}{3}$			$\frac{12}{4}$

27.

$\frac{11}{11}$			$\frac{12}{6}$
	3		
	2	$\frac{40}{10}$	
	4		$\frac{18}{6}$
			1

Beast Academy Practice 3D

PRACTICE | Solve each Fraction Link puzzle below.

28.

$\frac{42}{7}$				$\frac{15}{5}$
$\frac{36}{9}$			$\frac{60}{30}$	
			3	
			2	
	4			6

29.

5				
4	3		10	
		$\frac{80}{8}$	$\frac{45}{9}$	$\frac{21}{7}$
$\frac{32}{8}$				

30.

8				
2		$\frac{66}{11}$		
4		$\frac{30}{15}$		
	6			
			$\frac{20}{5}$	$\frac{56}{7}$

31.

$\frac{72}{8}$	7			3	
		9		5	
$\frac{63}{9}$	$\frac{75}{15}$	$\frac{30}{10}$			

32.

9	$\frac{16}{4}$			
	$\frac{50}{25}$			
		$\frac{54}{6}$	$\frac{18}{3}$	
6	2			4

33. ★

			$\frac{84}{12}$	$\frac{35}{7}$	$\frac{38}{19}$
	$\frac{24}{8}$				
5					
2			7		
	3				

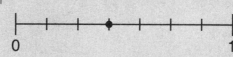

EXAMPLE | Label the number marked on the number line below.

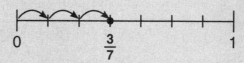

We can locate any fraction on the number line.

The tick marks split the number line between 0 and 1 into seven equal pieces. The length of each piece is $\frac{1}{7}$.

The dot marks the end of the third piece, so we count 3 lengths of $\frac{1}{7}$ from 0 to $\frac{3}{7}$.

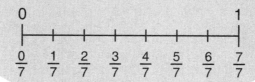

We can label all of the sevenths from 0 to 1 on the number line as shown below.

PRACTICE | Label the number marked on each number line below.

34.

35.

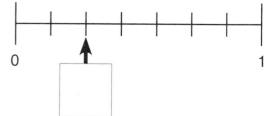

36.

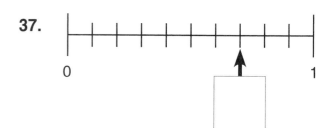

37.

38.

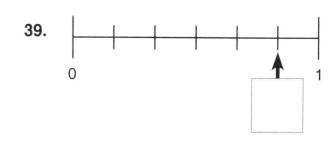

39.

EXAMPLE

Use a number line to compare the two fractions below. Then, place < or > in the circle to make a true statement.

$$\frac{2}{5} \bigcirc \frac{3}{7}$$

We can compare $\frac{2}{5}$ to $\frac{3}{7}$ on the number line.

To locate $\frac{2}{5}$, we divide the number line between 0 and 1 into five equal pieces and count 2 lengths of $\frac{1}{5}$.

To locate $\frac{3}{7}$, we divide the number line between 0 and 1 into seven equal pieces and count 3 lengths of $\frac{1}{7}$.

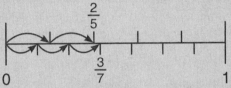

Since $\frac{2}{5}$ is to the left of $\frac{3}{7}$ on the number line, $\frac{2}{5}$ is less than $\frac{3}{7}$.

$$\frac{2}{5} \overset{<}{\bigcirc} \frac{3}{7}$$

Remember, the inequality symbols < and > always "eat" the larger number!

PRACTICE

Use the marked number lines to help you compare each pair of fractions. Then, place a <, >, or = in each circle below.

40. $\frac{3}{4} \bigcirc \frac{7}{9}$

41. $\frac{3}{11} \bigcirc \frac{1}{3}$

42. $\frac{6}{10} \bigcirc \frac{4}{7}$

43. $\frac{2}{6} \bigcirc \frac{4}{12}$

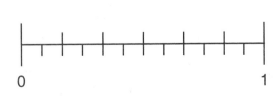

44. $\frac{8}{13} \bigcirc \frac{2}{3}$

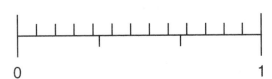

Fractions that represent the same number are equal and are called **equivalent** fractions.

EXAMPLE

Name two pairs of equivalent fractions on the number line below.

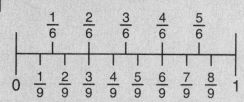

Equivalent fractions represent the same point on the number line. On the number line above, we see that $\frac{2}{6}$ and $\frac{3}{9}$ represent the same point, so they are equivalent fractions: $\frac{2}{6} = \frac{3}{9}$.

We also see that $\frac{4}{6}$ and $\frac{6}{9}$ represent the same point, so they are equivalent fractions: $\frac{4}{6} = \frac{6}{9}$.

We write our answer like this:

$$\frac{2}{6} = \frac{3}{9} \text{ and } \frac{4}{6} = \frac{6}{9}.$$

PRACTICE

Find all pairs of equivalent fractions between 0 and 1 marked on each number line below. The first number line has been labeled for you, and the first answer is given.

45.

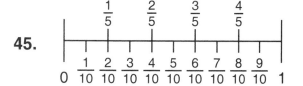

45. $\dfrac{1}{5} = \dfrac{2}{10}$ _____ _____

46.

46. _____ _____

47.

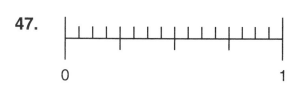

47. _____ _____ _____

Equivalent Fractions

EXAMPLE | Write two equivalent fractions for the point marked below.

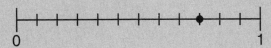

The number line between 0 and 1 is split into 12 equal pieces of length $\frac{1}{12}$. It takes 9 of these lengths to reach the mark indicated by the dot. So, the dot marks $\frac{9}{12}$.

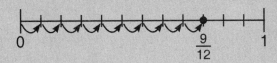

If we make every *third* tick mark bold, as shown below, then the number line between 0 and 1 is split into 4 equal pieces of length $\frac{1}{4}$. It takes 3 of these lengths to reach the mark indicated by the dot. So, the dot also marks $\frac{3}{4}$.

PRACTICE | Label each number marked below with two equivalent fractions.

48.

49.

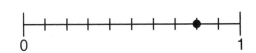

50.

51.

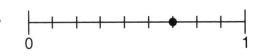

EXAMPLE | Locate $\frac{23}{6}$ on the number line below.

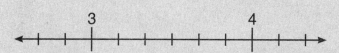

Since $18 \div 6 = 3$, we know that $\frac{18}{6} = 3$. Since $24 \div 6 = 4$, we know that $\frac{24}{6} = 4$. We label these fractions on the number line.

So, $\frac{23}{6}$ is between 3 and 4.

The tick marks divide the number line between 3 and 4 into 6 equal pieces, each with a length of $\frac{1}{6}$. We count five sixths past $\frac{18}{6}$ on the number line to mark $\frac{23}{6}$.

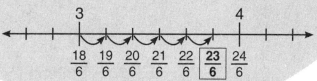

> We can use whole numbers to help us locate fractions on the number line.

PRACTICE

---

52. Label all of the missing fractions between 3 and 4 on the number line below.

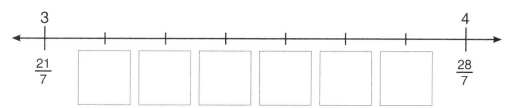

---

53. Label all of the missing fractions on the number line below.

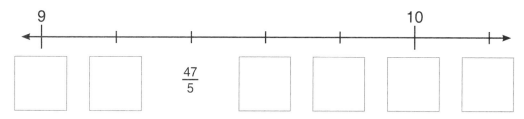

---

54. Label all of the missing fractions on the number line below.

PRACTICE | Locate and label each fraction listed below on the number line provided.

55. Locate and label $\frac{31}{5}$ on the number line below.

56. Locate and label $\frac{85}{9}$ on the number line below.

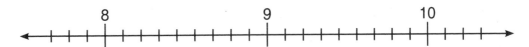

57. Locate and label $\frac{31}{4}$ on the number line below.

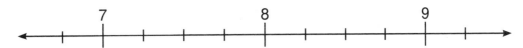

PRACTICE | Label all of the missing whole numbers on the number lines below.

58.

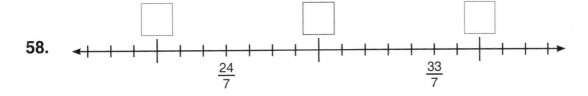

59.

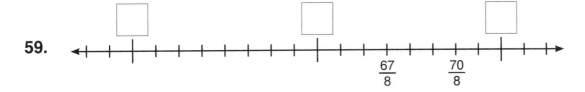

60. Label each number marked below with a fraction.

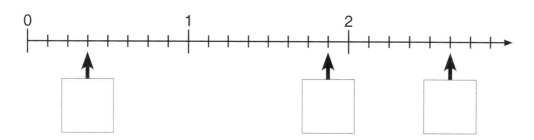

61. ★ Use the digits 1, 2, and 3 once each to write a fraction that is between 15 and 16.

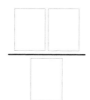

62. ★ Use the digits 4, 5, and 6 once each to write a fraction that is between 7 and 8.

63. ★ How many whole numbers are between $\frac{20}{3}$ and $\frac{40}{3}$?

63. _____

It's easier to tell how big a fraction is when it is written as a mixed number.

A *mixed number* is another way to write a fraction that is greater than 1. A mixed number is written as a whole number followed by a fraction that is less than 1.

For example, $3\frac{1}{6}$ is a mixed number that means the same thing as $3+\frac{1}{6}$. We read $3\frac{1}{6}$ as "three and one sixth."

On the number line below, we see that $3\frac{1}{6}$ equals $\frac{19}{6}$.

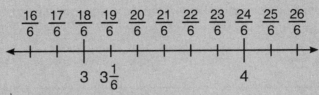

EXAMPLE | Write $\frac{43}{5}$ as a mixed number.

Since $40 \div 5 = 8$, we know that $\frac{40}{5} = 8$.

So, $\frac{43}{5}$ is three fifths more than 8.

Three fifths more than 8 is $8+\frac{3}{5}$, or $8\frac{3}{5}$.

So, $\frac{43}{5} = 8\frac{3}{5}$.

PRACTICE | Label each arrow on the number lines below with a mixed number.

64.

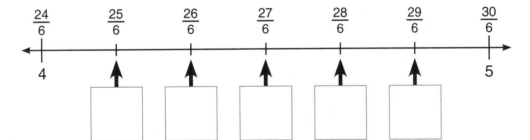

65.

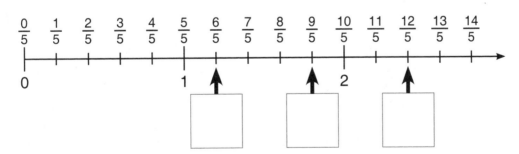

66.
★

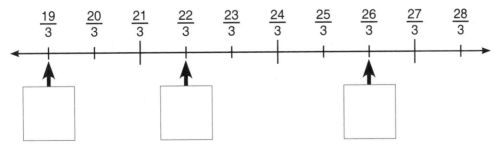

PRACTICE

67. Label the following mixed numbers on the number line below: $5\frac{3}{4}$, $6\frac{3}{4}$, and $7\frac{1}{4}$.

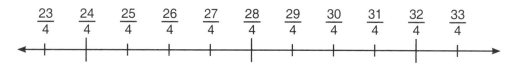

68. Write $\frac{19}{8}$ as a mixed number.

68. _____

69. Write $\frac{47}{6}$ as a mixed number.

69. _____

PRACTICE | Label each arrow with a mixed number on the number lines below.

70.

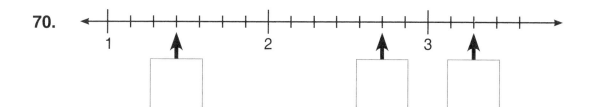

71.

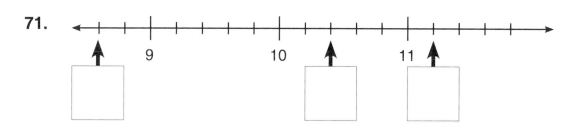

72. ★

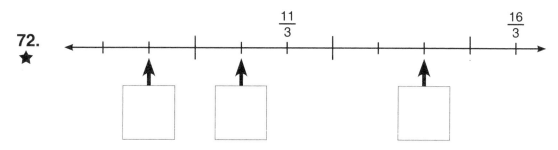

73. Some measurements for a recipe are given as fractions on the left.
Rewrite each fraction as a mixed number on the right.

$\frac{11}{4}$ cups flour $\qquad$ $\frac{3}{2}$ cups butter

$\frac{7}{4}$ cups chocolate chips $\qquad$ $\frac{5}{3}$ cups sugar

_____ cups flour $\qquad$ _____ cups butter

_____ cups chocolate chips $\qquad$ _____ cups sugar

PRACTICE | Label each number marked below with a *fraction*.

74.

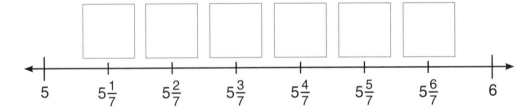

$5 \quad 5\frac{1}{7} \quad 5\frac{2}{7} \quad 5\frac{3}{7} \quad 5\frac{4}{7} \quad 5\frac{5}{7} \quad 5\frac{6}{7} \quad 6$

75.

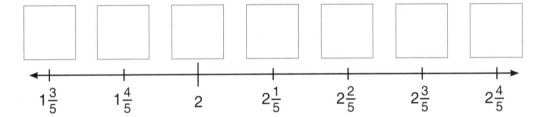

$1\frac{3}{5} \quad 1\frac{4}{5} \quad 2 \quad 2\frac{1}{5} \quad 2\frac{2}{5} \quad 2\frac{3}{5} \quad 2\frac{4}{5}$

76.

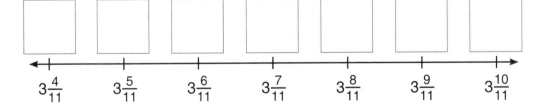

$3\frac{4}{11} \quad 3\frac{5}{11} \quad 3\frac{6}{11} \quad 3\frac{7}{11} \quad 3\frac{8}{11} \quad 3\frac{9}{11} \quad 3\frac{10}{11}$

77.

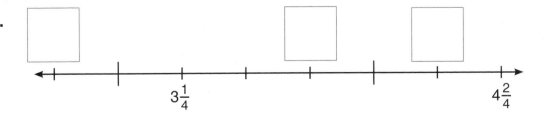

$3\frac{1}{4} \qquad\qquad\qquad 4\frac{2}{4}$

A fraction can be used to represent a part of something.

When we split something into parts, each part is a **fraction** of the whole.

EXAMPLE | What fraction of the circle below is shaded?

The circle is split into four *equal* pieces. So, each piece is $\frac{1}{4}$ of the whole circle. Since three pieces are shaded, $\frac{3}{4}$ of the circle is shaded.

PRACTICE | Each shape below has been split into equal parts. Write the fraction of each shape that is shaded.

78.

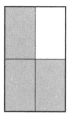

79.

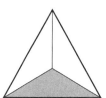

78. _____

79. _____

80.

81.

80. _____

81. _____

82.

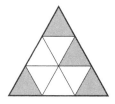

83.

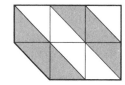

82. _____

83. _____

PRACTICE | Each shape below has been split into equal parts. Shade the given fraction of each shape below.

84. $\frac{3}{8}$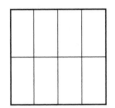

85. $\frac{7}{10}$

86. $\frac{2}{5}$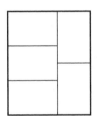

87. $\frac{5}{6}$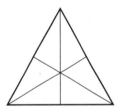

88. $\frac{4}{7}$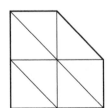

89. $\frac{5}{12}$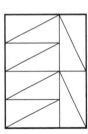

90. $\frac{2}{9}$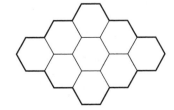

91. $\frac{5}{8}$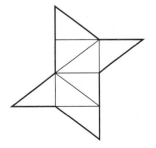

PRACTICE

92. Circle the equilateral triangle below that is $\frac{1}{4}$ shaded.

Explain why each other triangle is not $\frac{1}{4}$ shaded.

a.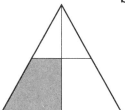

b.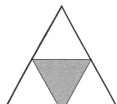

c.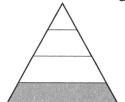

d.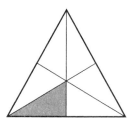

PRACTICE

Each shape below has been split into equal parts. Shade each shape below so that the given fraction is **unshaded**.

93. $\frac{2}{5}$ unshaded

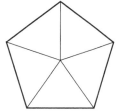

94. $\frac{3}{7}$ unshaded

95. $\frac{1}{6}$ unshaded

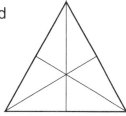

96. $\frac{6}{11}$ unshaded

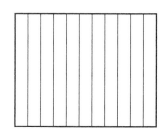

EXAMPLE | Shade $\frac{2}{3}$ of each circle below.

The circle on the left is split into 3 equal pieces. So, each piece is $\frac{1}{3}$ of the whole circle. To shade $\frac{2}{3}$ of the left circle, we shade 2 of these pieces:

> *Equivalent* fractions represent the same part of a whole.

The middle circle is split into 6 equal pieces. From 6 pieces, we can make three groups, each with 2 pieces. Each group is $\frac{1}{3}$ of the whole circle. So, to shade $\frac{2}{3}$ of the middle circle, we shade 4 pieces, which is $\frac{4}{6}$ of the circle.

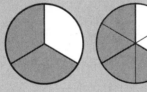

The right circle is split into 9 equal pieces. From 9 pieces, we can make three groups, each with 3 pieces. Each group is $\frac{1}{3}$ of the whole circle. To shade $\frac{2}{3}$ of the right circle, we shade 6 pieces, which is $\frac{6}{9}$ of the circle.

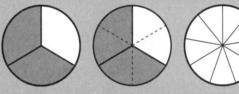

Since $\frac{2}{3}$, $\frac{4}{6}$, and $\frac{6}{9}$ represent the same part of identical circles, they are *equivalent* fractions: $\frac{2}{3} = \frac{4}{6} = \frac{6}{9}$.

PRACTICE | Shade $\frac{1}{4}$ of each rectangle below.

97.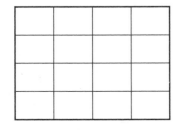

98.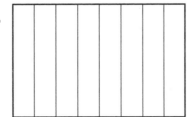

99.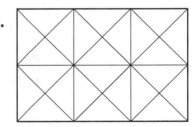

|

PRACTICE | Shade the given fraction of each shape below. Then, write an equivalent fraction to represent the part that you shaded.

100. $\frac{1}{2} = \underline{}$

101. $\frac{1}{5} = \underline{}$

102. $\frac{1}{3} = \underline{}$

103. $\frac{5}{6} = \underline{}$

104. $\frac{2}{5} = \underline{}$

105. $\frac{3}{4} = \underline{}$

106. Circle the shape below that is **not** $\frac{1}{4}$ shaded. Explain why each other shape is $\frac{1}{4}$ shaded.

a. 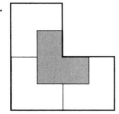   b. 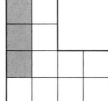   c.    d.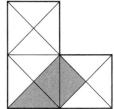

EXAMPLE | Find one fraction that is equivalent to $\frac{4}{5}$.

We mark $\frac{4}{5}$ on the number line.

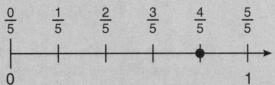

To find an equivalent fraction, we can split each piece of the number line above into two pieces, as shown.

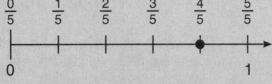

All together, the tick marks divide the number line between 0 and 1 into ten equal pieces, each with length $\frac{1}{10}$. It takes eight of these lengths to reach the mark indicated by the dot.

The dot also marks $\frac{8}{10}$. So, $\frac{4}{5}$ is **equivalent** to $\frac{8}{10}$.

Since we have doubled the number of pieces, this is the same as **multiplying** each of the numerator and denominator of $\frac{4}{5}$ by 2:

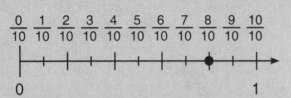

> Every fraction can be written in lots of ways!
>
> We can use multiplication to find **equivalent** fractions.

> Rewriting a fraction as an equivalent fraction is sometimes called **converting** the fraction.

PRACTICE | Convert each fraction below to an equivalent fraction by multiplying its numerator and denominator by the number given.

107.
$$\frac{4}{9} \overset{\times 2}{\underset{\times 2}{=}} \text{---}$$

108.
$$\frac{2}{7} \overset{\times 2}{\underset{\times 2}{=}} \text{---}$$

109.
$$\frac{4}{3} \overset{\times 4}{\underset{\times 4}{=}} \text{---}$$

110.
$$\frac{7}{10} \overset{\times 5}{\underset{\times 5}{=}} \text{---}$$

EXAMPLE | Find one fraction that is equivalent to $\frac{9}{12}$.

We mark $\frac{9}{12}$ on the number line.

Since 9 and 12 are both multiples of three, we can bold every *third* tick mark, as shown below.

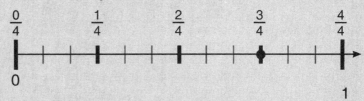

The bold tick marks divide the number line between 0 and 1 into four equal pieces of length $\frac{1}{4}$. It takes three of these lengths to reach the mark indicated by the dot.

Since the dot that marks $\frac{9}{12}$ also marks $\frac{3}{4}$, we have $\frac{9}{12} = \mathbf{\frac{3}{4}}$.

Bolding every third tick mark is the same as **dividing** each of the numerator and the denominator of $\frac{9}{12}$ by 3:

$$\frac{9}{12} \overset{\div 3}{\underset{\div 3}{=}} \frac{\mathbf{3}}{\mathbf{4}}$$

We can also use division to find equivalent fractions.

When we use division to find an equivalent fraction, this is called **simplifying** the fraction.

PRACTICE | Convert each fraction below to an equivalent fraction by dividing its numerator and denominator by the number given.

111. $\frac{6}{9} \overset{\div 3}{\underset{\div 3}{=}}$ —

112. $\frac{15}{25} \overset{\div 5}{\underset{\div 5}{=}}$ —

113. $\frac{36}{28} \overset{\div 4}{\underset{\div 4}{=}}$ —

114. $\frac{18}{81} \overset{\div 9}{\underset{\div 9}{=}}$ —

PRACTICE | Fill in the numerator or denominator that will make each equation below true.

115. $\frac{2}{3} = \frac{}{30}$

116. $\frac{4}{12} = \frac{1}{}$

117. $\frac{2}{14} = \frac{}{7}$

118. $\frac{5}{6} = \frac{}{12}$

119. $\frac{6}{7} = \frac{30}{}$

120. $\frac{12}{15} = \frac{}{5}$

121. $\frac{2}{8} = \frac{}{32}$

122. $\frac{6}{7} = \frac{}{42}$

123. $\frac{12}{78} = \frac{2}{}$

124. $\frac{13}{20} = \frac{}{80}$

PRACTICE

125. Five equivalent fractions are written below. Fill in the numerators and
★ denominators with the numbers that make the equation true.

$$\frac{2}{5} = \frac{}{10} = \frac{6}{} = \frac{30}{} = \frac{150}{}$$

126. Three equivalent fractions are written below. What are the values of
m and n?

$$\frac{1}{3} = \frac{4}{m} = \frac{m}{n}$$

126. $m = $ _____

$n = $ _____

127. If s is a whole number, then what is the value of s?
★

$$\frac{1}{s} = \frac{s}{16}$$

127. $s = $ _____

128. Two pairs of equivalent fractions are written below. What is x?
★
★

$$\frac{3}{a} = \frac{9}{b} \quad \text{and} \quad \frac{5}{a} = \frac{x}{b}$$

128. $x = $ _____

PRACTICE

129. Adding 2 to the numerator and denominator of $\frac{1}{4}$ gives us $\frac{3}{6}$.
Is $\frac{3}{6}$ equivalent to $\frac{1}{4}$?

129. _____

130. Does adding the same number to the numerator and denominator
of a fraction always create an equivalent fraction? Explain.

131. Subtracting 2 from the numerator and denominator of $\frac{6}{5}$ gives us $\frac{4}{3}$.
Is $\frac{4}{3}$ equivalent to $\frac{6}{5}$?

131. _____

132. Does subtracting the same number from the numerator and
denominator of a fraction always create an equivalent fraction?
Explain.

A fraction is in **simplest form** when 1 is the only whole number that divides both the numerator and denominator without a remainder.

EXAMPLE | Write $\frac{30}{48}$ in simplest form.

30 and 48 are both multiples of 2, so

$$\frac{30}{48} \overset{\div 2}{\underset{\div 2}{=}} \frac{15}{24}$$

Then, $\frac{15}{24}$ can be simplified further.
15 and 24 are both multiples of 3, so

$$\frac{30}{48} = \frac{15}{24} \overset{\div 3}{\underset{\div 3}{=}} \frac{5}{8}$$

Since 1 is the only whole number that divides each of 5 and 8 with no remainder, $\frac{5}{8}$ is in simplest form.

— *or* —

We could simplify $\frac{30}{48}$ in one step by dividing each of 30 and 48 by 6:

$$\frac{30}{48} \overset{\div 6}{\underset{\div 6}{=}} \frac{5}{8}$$

PRACTICE | Simplify each fraction below until it is in simplest form.

133. $\frac{6}{8} =$

134. $\frac{9}{18} =$

135. $\frac{15}{45} =$

136. $\frac{22}{40} =$

137. $\frac{12}{18} =$

138. $\frac{13}{24} =$

139. $\frac{32}{56} =$

140. $\frac{72}{60} =$

PRACTICE | Label each point marked below as a fraction in simplest form.

141.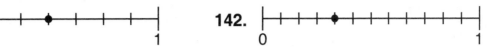

142.

143.

144.

145.

146.

PRACTICE | Each shape below has been split into equal parts.
Write the **shaded** fraction of each shape in simplest form.

147. _____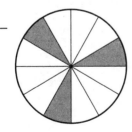

148. _____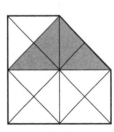

149. _____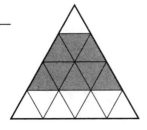

150. _____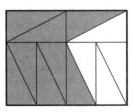

In a **Constellation Puzzle**, the goal is to connect every group of equivalent fractions with a path of straight lines. Below is an example of a Constellation Puzzle and its solution:

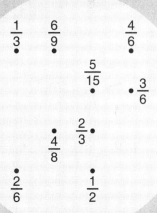
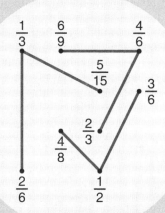

In this puzzle, we have $\frac{5}{15}=\frac{1}{3}=\frac{2}{6}$, and $\frac{6}{9}=\frac{2}{3}=\frac{4}{6}$, as well as $\frac{4}{8}=\frac{1}{2}=\frac{3}{6}$. Each set of equivalent fractions is connected by a path using the following rules:

1. Dots on the paths must be connected by straight lines.

2. Paths may not cross.

3. When connecting two dots, you may not draw a line through a third dot.

PRACTICE | Solve each Constellation Puzzle below.

151.

$\frac{4}{12}$• •$\frac{3}{6}$ $\frac{8}{24}$• •$\frac{1}{2}$

•$\frac{2}{4}$

•$\frac{1}{3}$

152.

$\frac{1}{7}$• •$\frac{9}{63}$

•$\frac{5}{25}$

$\frac{8}{40}$• $\frac{5}{35}$•

•$\frac{1}{5}$

PRACTICE | Solve each Constellation Puzzle below.

153.

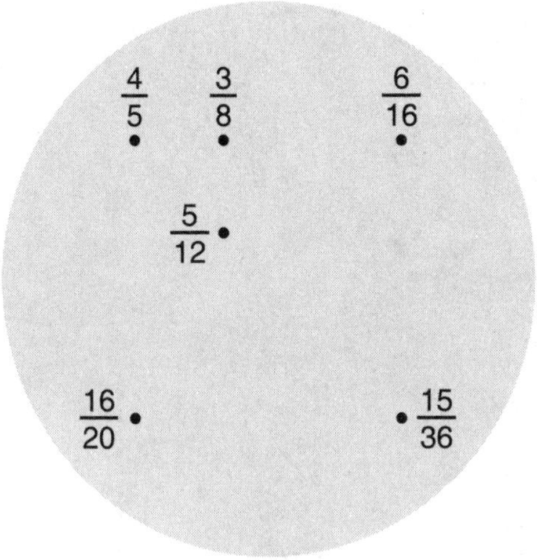

$\frac{4}{5}$ $\frac{3}{8}$ $\frac{6}{16}$

$\frac{5}{12}$

$\frac{16}{20}$ $\frac{15}{36}$

154.

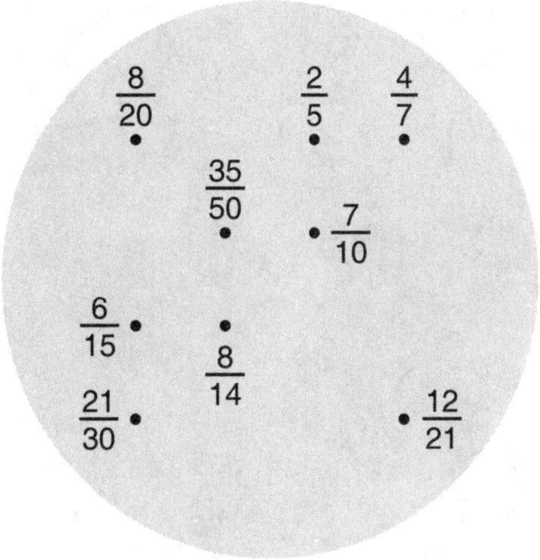

$\frac{8}{20}$ $\frac{2}{5}$ $\frac{4}{7}$

$\frac{35}{50}$ $\frac{7}{10}$

$\frac{6}{15}$ $\frac{8}{14}$

$\frac{21}{30}$ $\frac{12}{21}$

155. ★

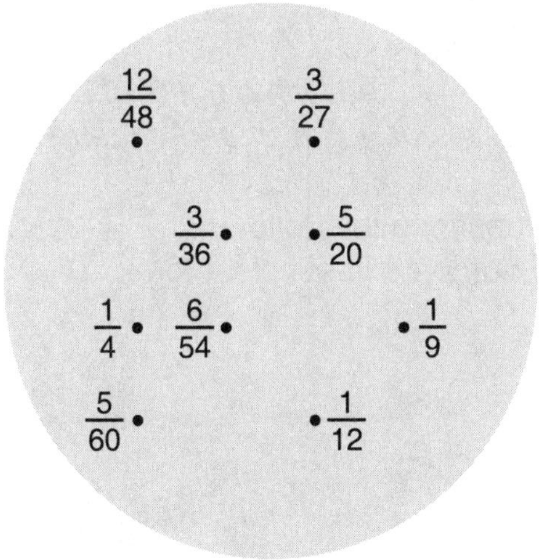

$\frac{12}{48}$ $\frac{3}{27}$

$\frac{3}{36}$ $\frac{5}{20}$

$\frac{1}{4}$ $\frac{6}{54}$ $\frac{1}{9}$

$\frac{5}{60}$ $\frac{1}{12}$

156. ★

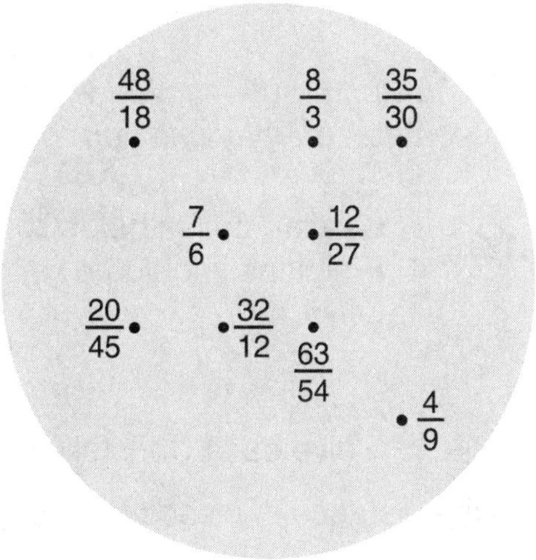

$\frac{48}{18}$ $\frac{8}{3}$ $\frac{35}{30}$

$\frac{7}{6}$ $\frac{12}{27}$

$\frac{20}{45}$ $\frac{32}{12}$ $\frac{63}{54}$

$\frac{4}{9}$

157. ★

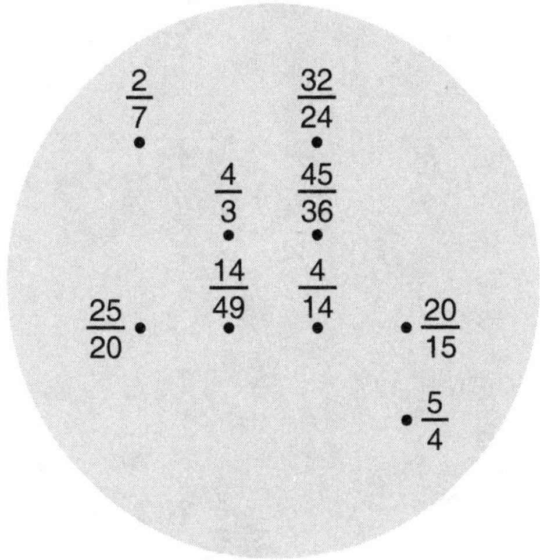

$\frac{2}{7}$ $\frac{32}{24}$

$\frac{4}{3}$ $\frac{45}{36}$

$\frac{14}{49}$ $\frac{4}{14}$

$\frac{25}{20}$ $\frac{20}{15}$

$\frac{5}{4}$

158. ★

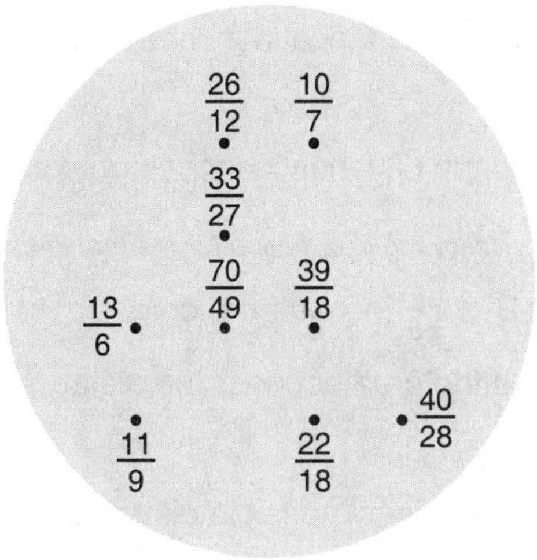

$\frac{26}{12}$ $\frac{10}{7}$

$\frac{33}{27}$

$\frac{70}{49}$ $\frac{39}{18}$

$\frac{13}{6}$

$\frac{40}{28}$

$\frac{11}{9}$ $\frac{22}{18}$

A *mixed number in simplest form* is a mixed number whose fractional part is in simplest form.

Most rulers can be used to measure fractions of an inch.

On the ruler below, the large tick marks are numbered. These marks represent whole inches. Each inch is split into eighths of an inch.

Measurements are given as *mixed numbers in simplest form*.

For example, the crayon below is 2 whole inches long, plus 6 eighths $\left(\frac{6}{8}\right)$ of an inch. We can simplify $\frac{6}{8}$ to $\frac{3}{4}$.

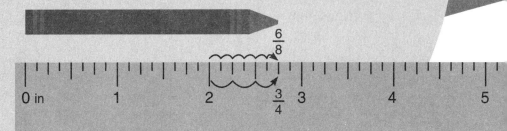

So, the crayon is $2+\frac{3}{4}$ inches long.
We write this as $2\frac{3}{4}$ **inches**.

For the following problems, you will need a ruler that is marked with eighths of an inch like the one above. You may use your own or print a ruler to cut out at BeastAcademy.com.

PRACTICE | Use the points below to answer the questions that follow.
Fractions should be written as mixed numbers in simplest form.

•
A

•
B

•
C

159. What is the distance in inches from point A to point B?

159. _____

160. What is the distance in inches from point B to point C?

160. _____

161. What is the distance in inches from point A to point C?

161. _____

Use the following information for the problems below:

Point D is $1\frac{1}{8}$ inches from point B.

162. ★ What is the greatest possible distance from point A to point D?

162. _____

163. ★ What is the least possible distance from point A to point D?

163. _____

PRACTICE | Bobby the bacterium lives in the tiny house marked with a B. His best friend, Carl, lives in the house marked with a C. Use the labeled houses to answer the questions that follow.

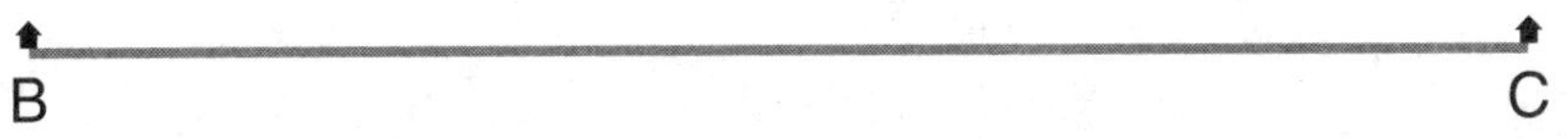

B _____ C

164. What is the distance in inches from Bobby's house to Carl's house?

164. _____

165. On Monday, Bobby begins walking to Carl's. He walks $1\frac{1}{8}$ inches from his house before stopping to take a break. How far is Bobby from Carl's house when he stops to take a break?

165. _____

166. After his break, Bobby walks $2\frac{5}{8}$ more inches towards Carl's house. All together, how far has Bobby walked from his house?

166. _____

167. ★ On Tuesday, Bobby walks from his house to Carl's and back. How far does Bobby walk on Tuesday?

167. _____

168. ★★ On Wednesday, while walking to Carl's house, Bobby stops halfway between the two houses to tie his shoe. How far is Bobby from home when he stops to tie his shoe?

168. _____

Fractions are another way to write division. So, we can answer a division problem as a fraction or mixed number, instead of a quotient and remainder. We typically write the answer in **simplest form**.

EXAMPLE

Compute $15 \div 6$. Write the answer as a mixed number in simplest form.

First, we can write $15 \div 6$ as $\frac{15}{6}$.

We can write $\frac{15}{6}$ as a mixed number: $2\frac{3}{6}$.

Then, $\frac{3}{6}$ can be simplified to $\frac{1}{2}$. So, $15 \div 6 = 2\frac{1}{2}$.

— *or* —

We can simplify $\frac{15}{6}$ to $\frac{5}{2}$. Then, we can write $\frac{5}{2}$ as a mixed number: $2\frac{1}{2}$. Since $2\frac{1}{2}$ is in simplest form, we have $15 \div 6 = 2\frac{1}{2}$.

PRACTICE

Give the answer to each division problem below as a **mixed number in simplest form**.

169. $7 \div 2 =$ _____

170. $14 \div 9 =$ _____

171. $26 \div 8 =$ _____

172. $86 \div 10 =$ _____

173. What is $4 \div 12$? Write your answer as a fraction in simplest form.

173. _____

174. What is $36 \div 84$? Write your answer as a fraction in simplest form.

174. _____

175. The perimeter of a square is 14 inches. What is the side length of the square?

175. _____

176. Four small triangles are attached as shown below to make a larger triangle with a perimeter of 8 cm. What is the length in centimeters of one side of a small triangle?

176. _____

It is easiest to compare fractions when the numerators or the denominators are the same.
To compare fractions, it can be useful to convert one or both fractions.

EXAMPLE | Which number is greater, $\frac{3}{5}$ or $\frac{8}{15}$?

We can compare $\frac{3}{5}$ to $\frac{8}{15}$ by converting $\frac{3}{5}$ to an equivalent fraction with denominator 15. To do this, we multiply the numerator and denominator of $\frac{3}{5}$ by 3:

$$\frac{3}{5} \overset{\times 3}{\underset{\times 3}{=}} \frac{9}{15}$$

$\frac{9}{15}$ is greater than $\frac{8}{15}$.

So, $\frac{3}{5}$ **is greater than** $\frac{8}{15}$.

We can also write $\frac{3}{5}$ ⊗ $\frac{8}{15}$.

PRACTICE | Place $<$ or $>$ in each circle to compare each pair of fractions below.

177. $\frac{4}{7} \bigcirc \frac{5}{7}$

178. $\frac{1}{4} \bigcirc \frac{3}{8}$

179. $\frac{3}{4} \bigcirc \frac{11}{16}$

180. $\frac{2}{3} \bigcirc \frac{5}{6}$

181. $\frac{9}{16} \bigcirc \frac{5}{8}$

182. $\frac{10}{21} \bigcirc \frac{3}{7}$

When comparing fractions, we usually convert them so that their **denominators** are the same.

Sometimes, it is easier to compare two fractions by converting them so that their **numerators** are the same.

EXAMPLE | Which number is greater, $\frac{4}{7}$ or $\frac{2}{5}$?

We can compare $\frac{4}{7}$ to $\frac{2}{5}$ by converting $\frac{2}{5}$ to an equivalent fraction with numerator 4.

To do this, we multiply the numerator and denominator of $\frac{2}{5}$ by 2:

$$\frac{2}{5} \overset{\times 2}{\underset{\times 2}{=}} \frac{4}{10}$$

Sevenths are larger than tenths.

So, $\frac{4}{7}$ is greater than $\frac{4}{10}$, and $\frac{4}{7}$ **is greater than** $\frac{2}{5}$.

We can also write $\frac{4}{7} \gt \frac{2}{5}$.

PRACTICE | Place a < or > in each circle to compare each pair of fractions below.

183. $\frac{5}{8} \bigcirc \frac{5}{11}$

184. $\frac{2}{7} \bigcirc \frac{4}{13}$

185. $\frac{3}{5} \bigcirc \frac{6}{11}$

186. $\frac{4}{9} \bigcirc \frac{20}{43}$

187. $\frac{9}{31} \bigcirc \frac{3}{11}$

188. $\frac{24}{43} \bigcirc \frac{4}{7}$

PRACTICE

189. Circle all of the fractions below that are greater than $\frac{1}{2}$.

$$\frac{2}{6} \qquad \frac{4}{6} \qquad \frac{3}{8} \qquad \frac{5}{8} \qquad \frac{4}{10} \qquad \frac{6}{10} \qquad \frac{5}{12} \qquad \frac{7}{12}$$

190. Circle all of the fractions below that are less than $\frac{1}{2}$.

$$\frac{3}{5} \qquad \frac{3}{7} \qquad \frac{4}{7} \qquad \frac{4}{9} \qquad \frac{5}{9} \qquad \frac{5}{11} \qquad \frac{6}{11} \qquad \frac{6}{13}$$

191. Order the following fractions from least to greatest: 191. _____ _____ _____

$$\frac{4}{9} \qquad \frac{2}{3} \qquad \frac{8}{15}$$

PRACTICE | Use the given numbers to fill in the missing numerators and denominators so that the fractions are in order from least to greatest. Every fraction should be **less than 1** and in **simplest form**.

192. 6, 7, 8: $\dfrac{5}{} < \dfrac{5}{} < \dfrac{5}{}$

193. 7, 8, 9: $\dfrac{4}{} < \dfrac{5}{} < \dfrac{6}{}$
★

194. 3, 7, 11: $\dfrac{3}{} < \dfrac{}{7} < \dfrac{5}{}$
★

195. 3, 4, 5, 6: $\dfrac{}{7} < \dfrac{2}{} < \dfrac{}{}$
★
★

PRACTICE

196. Alex ate $\frac{1}{2}$ of a cake. Grogg ate $\frac{2}{3}$ of a pie. Can you tell who ate more? Why or why not?

197. Use the four fractions below to label the four marked points on the number line.

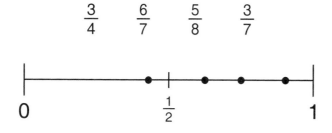

$$\frac{3}{4} \qquad \frac{6}{7} \qquad \frac{5}{8} \qquad \frac{3}{7}$$

198. What whole-number value of y makes the following true?

198. $y =$ _____

$$\frac{2}{11} < \frac{1}{y} < \frac{2}{9}$$

PRACTICE

199. ★ Place <, >, or = in the circle below to compare the pair of fractions.

$$\frac{2}{5} \bigcirc \frac{7}{21}$$

200. ★ Place <, >, or = in the circle below to compare the pair of fractions.

$$\frac{37}{99} \bigcirc \frac{73}{200}$$

201. ★ If $\frac{1}{z}$ is a unit fraction greater than $\frac{2}{7}$ and less than $\frac{3}{8}$, then what is the value of z?

201. $z =$ _____

202. ★★ Lizzie writes a number that is between $\frac{1}{10}$ and $\frac{1}{9}$. She only uses the digits 7, 8, and 9 in the numerator and denominator, and each digit is used exactly once. Find the **two** numbers Lizzie could have written.

202. _____ or _____

PRACTICE | Using a straightedge, split each shape below into equal parts so that you can then shade the given fraction of each shape.

203. $\frac{1}{2}$

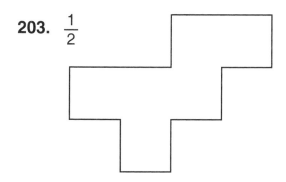

204. $\frac{2}{5}$

205. $\frac{3}{8}$

206. $\frac{4}{11}$

PRACTICE | Using a straightedge, split each shape below into equal parts so that you can then shade the given fraction of each shape.

207. $\frac{5}{6}$

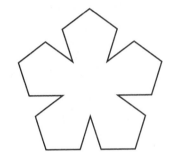

208. $\frac{4}{9}$

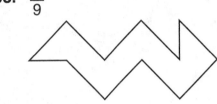

209. $\frac{7}{10}$

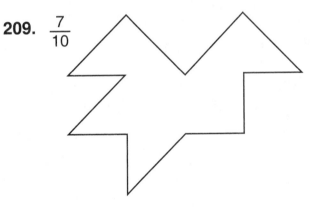

210. ★ $\frac{3}{4}$

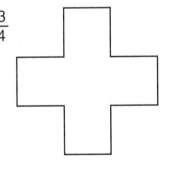

211. ★ $\frac{1}{3}$

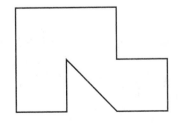

212. ★ $\frac{5}{7}$

CHAPTER 11
Estimation

Use this Practice book with
Guide 3D from BeastAcademy.com.

Recommended Sequence:

Book	Pages:
Guide:	52–66
Practice:	47–61
Guide:	67–73
Practice:	62–75

You may also read the entire chapter
in the Guide before beginning the
Practice chapter.

When we **estimate**, we make a thoughtful guess.

Sometimes, an exact answer to a problem is nearly impossible to find, or an answer that is close is all you need.

When an exact answer is not necessary, you can **estimate**.

PRACTICE For each example below, decide whether an estimate or an exact amount is more appropriate.
Write **exact** or **estimate** in the blank provided.

1. The number of hairs you have on your head.

1. _____

2. The number of students on the bus ride back from a field trip.

2. _____

3. The distance from Earth to the Moon, in miles.

3. _____

4. The number of points scored by each team in a basketball game.

4. _____

5. The width of a door to be installed by a carpenter.

5. _____

6. The number of gallons of water that flow down the Mississippi River each minute.

6. _____

7. The height of the bar in the Beast Olympics high jump event.

7. _____

8. The number of apples on all of the trees in an orchard.

8. _____

Sometimes we estimate by **rounding**. We can round whole numbers to the nearest ten, hundred, thousand, or any other place value!

EXAMPLES | Round each of 743 and 175 to the nearest ten.

Since 743 is closer to 740 than to 750, we round 743 **down** to **740**.

Since 175 is exactly between 170 and 180, it is not closer to 170 or 180. Numbers in the middle are rounded up. So, 175 **rounds up** to **180**.

Because numbers in the middle are rounded up, we only need to look at the digit to the right of the place value that we are rounding to.

If the digit is a 5, 6, 7, 8, or 9, we round **up**.
If the digit is a 0, 1, 2, 3, or 4, we round **down**.

PRACTICE | Round each number below to the **nearest ten**.

9. 62 rounds to _____

10. 486 rounds to _____

11. 101 rounds to _____

12. 1,395 rounds to _____

PRACTICE | Round each number below to the **nearest hundred**.

13. 1,283 rounds to _____

14. 5,135 rounds to _____

15. 44,445 rounds to _____

16. 199,999 rounds to _____

PRACTICE | Round each number below to the *nearest thousand*.

17. 7,890 rounds to _____

18. 45,678 rounds to _____

19. 2,000,499 rounds to _____

20. 99,615 rounds to _____

PRACTICE | Solve each rounding problem below.

21. When rounded to the nearest hundred, Bill's favorite number rounds to 300. What is the smallest possible value of Bill's favorite number?

21. _____

22. What is the largest whole number that rounds to 6,000 when rounded to the nearest thousand?

22. _____

23. How many different whole numbers round to 50 when rounded to the nearest ten?

23. _____

PRACTICE | Solve each rounding problem below.

24. Kim rounds 777 to the nearest ten. Jim rounds 777 to the nearest hundred. How much larger is Jim's estimate than Kim's?

24. _____

25. When Kyle rounds his favorite 2-digit number to the nearest hundred, his estimate is 18 more than the original number. What is Kyle's favorite 2-digit number?

25. _____

26. Kelly's favorite number is between 582 and 761. When she rounds her favorite number to the nearest hundred, her estimate is 36 more than the original number. What is Kelly's favorite number?

26. _____

27. ★ Kina picks a number. Whether Kina rounds her number to the nearest ten, hundred, or thousand, she always gets 8,000. What is the smallest number that Kina could have picked?

27. _____

28. ★ How many different whole numbers round to 400 when rounded to the nearest hundred **and** round to 450 when rounded to the nearest ten?

28. _____

29. ★★ Winnie picks a whole number. Grogg rounds Winnie's number to the nearest ten. Then, Lizzie rounds Grogg's number to the nearest hundred. Finally, Alex rounds Lizzie's number to the nearest thousand and gets 9,000. What is the smallest number that Winnie could have chosen?

29. _____

If the numerator of a fraction is **less** than half its denominator, then the fraction is less than one half.

If the numerator of a fraction is **more** than half its denominator, then the fraction is more than one half.

EXAMPLE | Round $6\frac{5}{8}$ to the nearest whole number.

To round a mixed number to the nearest whole number, we look at its fractional part.

If the fractional part is less than one half, we **round down**.

If the fractional part is one half or more, we **round up**.

The fractional part of $6\frac{5}{8}$ is $\frac{5}{8}$. Since 5 is more than half of 8, we know that $\frac{5}{8}$ is greater than $\frac{1}{2}$.

Therefore, we round $6\frac{5}{8}$ up to **7**.

We can write $6\frac{5}{8} \approx 7$.

The symbol ≈ means "is approximately" or "is close to."

PRACTICE | Round each fraction or mixed number below to the **nearest whole number**.

30. $12\frac{1}{3} \approx$ _____

31. $15\frac{3}{4} \approx$ _____

32. $23\frac{5}{9} \approx$ _____

33. $\frac{9}{7} \approx$ _____

34. $\frac{19}{5} \approx$ _____

35. $\frac{35}{8} \approx$ _____

36. $29\frac{50}{99} \approx$ _____

37. $\frac{87}{11} \approx$ _____

EXAMPLE | Estimate the product of 72 and 287.

We can use rounded values to compute estimates!

Since 72 is about 70 and 287 is about 300, we expect 72×287 to be about 70×300.

Since 70×300 is 21,000, we estimate that
$$72 \times 287 \approx 70 \times 300 = \textbf{21,000}.$$

In fact, $72 \times 287 = 20,664$. So, our estimate differs from the actual value by $21,000 - 20,664 = 336$.

Even though 336 may seem like a big number, compared to 21,000, the number 336 is pretty small. So, 21,000 is a good estimate of 72×287.

A good estimate is easy to compute **and** is close to the actual answer.

PRACTICE | Solve each problem below by estimating.

38. Is the sum $5,796 + 9,359$ closer to 15,000 or to 150,000?

38. _____

39. Is the product $1,123 \times 47$ closer to 5,000 or to 50,000?

39. _____

40. Is the sum $8\frac{3}{4} + 13\frac{1}{3}$ closer to 20 or to 30?

40. _____

41. How many digits long is the sum $72,345 + 68,922$?

41. _____

PRACTICE | Solve each problem below by estimating.

42. Circle the number below that is equal to 37×27 without computing the exact answer.

  555   999   6,789   9,999

43. To estimate the product 89×203, Grogg rounds both numbers to the nearest hundred, then multiplies. Alex rounds both numbers to the nearest ten before multiplying. How much larger is Grogg's estimate than Alex's?

43. _____

44. Circle the number below that is equal to 231×533 without computing the exact answer.

  1,234   12,321   123,123   1,234,567

45. The quotient $24\frac{1}{3} \div 7\frac{8}{9}$ is close to what whole number?

45. _____

46. Adam estimates $6 \times 704 \approx 10 \times 704 = 7,040$.
Jon estimates $6 \times 704 \approx 6 \times 700 = 4,200$.
Can you tell which estimate is closer without computing 6×704?
Explain.

47. Globb estimates the product $49 \times 499 \times 4,999$ by rounding all three numbers to the nearest hundred, then multiplying. What estimate does he get? Is this a ***good*** estimate? If so, explain why.
If not, how should Globb have rounded instead?

ESTIMATION
Computing Estimates

An answer is **reasonable** if it makes sense.

By using estimation to check whether our answers are reasonable, we can often catch mistakes and correct them.

EXAMPLE | Which of the four computations below is **not** reasonable?

$681+225 = 906$ $44\times73 = 3,212$

$75\times18 = 1,350$ $897+6,041 = 15,011$

We can estimate the value of each sum or product, then compare our estimates to the values above.

$681+225 \approx 700+200 = 900$, which is very close to 906.
$44\times73 \approx 40\times70 = 2,800$, which is pretty close to 3,212.
$75\times18 \approx 80\times20 = 1,600$, which is pretty close to 1,350.
$897+6,041 \approx 900+6,000 = 6,900$, which is less than
      half of 15,011.

Our estimate suggests that **$897+6,041 = 15,011$** is not reasonable. A mistake must have been made in the computation.

In fact, $897+6,041 = 6,938$.

PRACTICE | Place a ✓ in the box next to every computation that is reasonable. Place a ✗ next to every computation that is not reasonable. For each, write an estimate that shows why you chose to write a ✓ or ✗.

48. ☐ $79\times107 = 8,453$ **49.** ☐ $1,297+680 = 8,097$ **50.** ☐ $67\times1,008 = 7,236$

51. ☐ $532+118 = 650$ **52.** ☐ $89+778 = 1,668$ **53.** ☐ $603\times90 = 54,270$

54. ☐ $7\times986 = 69,020$ **55.** ☐ $78\times215 = 1,677$ **56.** ☐ $977+14,040 = 15,017$

57. ☐ $9\times11\times8 = 792$ **58.** ☐ $866+1,671 = 2,537$ **59.** ☐ $7\times(89+23) = 7,784$

In a Short Circuit puzzle, each dot is labeled with an expression. The goal is to draw wires that connect each labeled dot on the left to a dot labeled with an equal expression on the right. The wires must not leave the room, cross each other, or pass through walls.

EXAMPLE | Complete the Short Circuit puzzle below by connecting the three pairs of equal expressions.

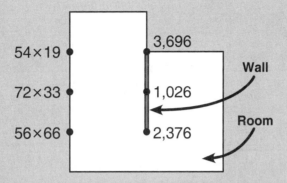

Since the numbers on the right side of the puzzle are not close in value, we do not need to know the exact values of the products on the left to match them to the numbers on the right. So, we begin by estimating the value of each product on the left.

54×19 is approximately 50×20 = 1,000. So, we guess that 54×19 = 1,026.
72×33 is approximately 70×30 = 2,100. So, we guess that 72×33 = 2,376.
56×66 is approximately 60×70 = 4,200. So, we guess that 56×66 = 3,696.

We must connect 54×19 to the dot labeled 1,026, 72×33 to the dot labeled 2,376, and 56×66 to 3,696.

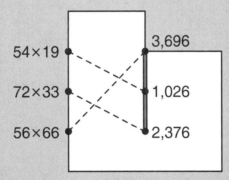

To avoid crossing wires, we can connect 56×66 to 3,696 by going around the wall in the middle of the room as shown below. We connect the other pairs with straight lines.

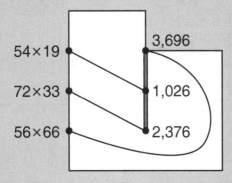

PRACTICE | Complete each Short Circuit puzzle below.
We recommend you use a pencil.

60.

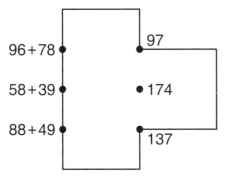

96+78
58+39
88+49

97
174
137

61.

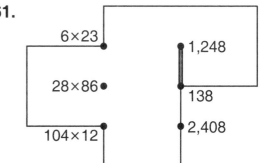

6×23
28×86
104×12

1,248
138
2,408

62.

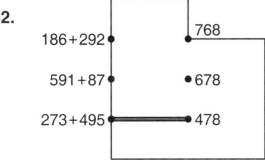

186+292
591+87
273+495

768
678
478

63.

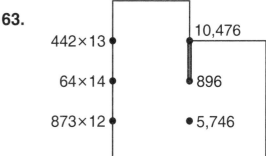

442×13
64×14
873×12

10,476
896
5,746

64.

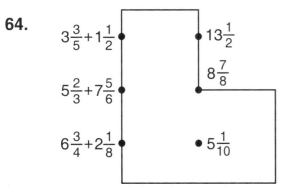

$3\frac{3}{5}+1\frac{1}{2}$
$5\frac{2}{3}+7\frac{5}{6}$
$6\frac{3}{4}+2\frac{1}{8}$

$13\frac{1}{2}$
$8\frac{7}{8}$
$5\frac{1}{10}$

65.

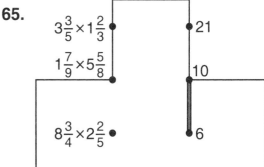

$3\frac{3}{5}\times1\frac{2}{3}$
$1\frac{7}{9}\times5\frac{5}{8}$
$8\frac{3}{4}\times2\frac{2}{5}$

21
10
6

PRACTICE | Complete each Short Circuit puzzle below.
We recommend you use a pencil.

66.
★

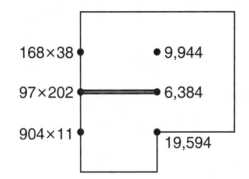

67.
★

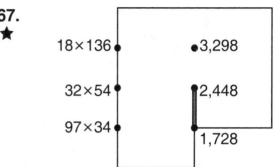

68.
★

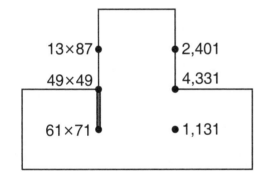

69.
★

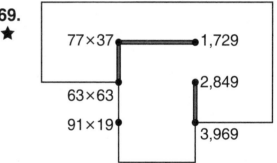

70.
★
★

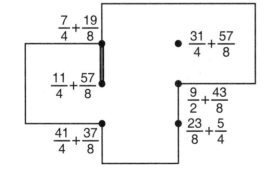

71.
★
★

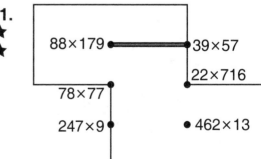

It can be helpful to know whether an estimate is greater than or less than the exact value.

An estimate that is larger than the actual value is called an **overestimate**. An estimate that is smaller than the actual value is called an **underestimate**.

EXAMPLE

State whether each of the following is an **over**estimate or an **under**estimate.

a) $275 \times 492 \approx 300 \times 500 = 150{,}000$.

b) $5\frac{1}{9} + 3\frac{1}{3} \approx 5 + 3 = 8$.

a) Both numbers were rounded up and then multiplied. This makes the estimate larger than the real answer. So, 150,000 is an **overestimate**.

The actual value of 275×492 is 135,300.

b) Both numbers were rounded down and then added. This makes the estimate smaller than the real answer. So, 8 is an **underestimate**.

The actual value of $5\frac{1}{9} + 3\frac{1}{3}$ is $8\frac{4}{9}$.

PRACTICE | Answer each question about overestimates and underestimates below.

72. Is $11\frac{17}{20} + 14\frac{4}{7}$ greater than or less than $12 + 15 = 27$?

72. _____

73. Is 23×31 greater than or less than $20 \times 30 = 600$?

73. _____

74. Kevin estimates $13\frac{2}{9} \times 2\frac{3}{8}$ by computing $13 \times 2 = 26$. Is 26 an overestimate or an underestimate?

74. _____

75. Pongo estimates $5{,}725 + 18{,}842$ by computing $6{,}000 + 19{,}000 = 25{,}000$. Is 25,000 an overestimate or an underestimate?

75. _____

76. ★ ✏ When estimating a product of two numbers, Norton rounds one of the numbers up and rounds the other number down. Can you tell if Norton's estimate is an overestimate or an underestimate without more information? Explain.

PRACTICE | Fill each circle below with < or > to indicate which expression is greater without computing the actual sums and products.

77. $85 \times 17 \bigcirc 90 \times 20$

78. $8{,}519 + 4{,}672 \bigcirc 9{,}000 + 5{,}000$

79. $461 \times 11 \bigcirc 460 \times 10$

80. $97 \times 9 \bigcirc 1{,}000$

81. $51\frac{5}{13} + 16\frac{1}{8} \bigcirc 67$

82. $7\frac{3}{4} \times 6\frac{7}{9} \bigcirc 56$

PRACTICE | Answer each question about overestimates and underestimates below.

83. There are 12 inches in a foot and 5,280 feet in a mile. Which distance is greater: 1 mile or 50,000 inches?

83. _____

84. ★ A sheet of cookies holds 36 cookies. Lunch Lady Lydia bakes 21 sheets of cookies. Will this be enough to give each of 700 students at least one cookie?

84. _____

85. ★ 🖉 Rosencrantz and Guildenstern use square paving blocks to make a garden patio. The rectangular patio will be 9 blocks wide and 13 blocks long. Each block weighs 23 pounds. A truck can deliver up to 2,000 pounds of blocks. Can one truck deliver enough blocks to build the patio? Explain.

EXAMPLE

State whether each of the following is an **over**estimate or an **under**estimate.

a) $624 - 264 \approx 600 - 300 = 300$.

b) $782 - 622 \approx 800 - 600 = 200$.

We can think of subtraction as finding the distance between two numbers on the number line.

a) 600 is less than 624, and 300 is greater than 264. So, 624 and 264 are farther apart than 600 and 300.

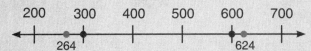

This means that $600 - 300 = 300$ is an **underestimate**.
In fact, $624 - 264 = 360$, which is greater than 300.

b) 800 is greater than 782, and 600 is less than 622. So, 782 and 622 are closer together than 800 and 600.

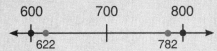

This means that $800 - 600 = 200$ is an **overestimate**.
In fact, $782 - 622 = 160$, which is less than 200.

PRACTICE | Solve each estimation problem below by writing "greater" or "less" in the blank provided.

86. Is $501 - 299$ greater than or less than $500 - 300 = 200$?

86. _____

87. Is $12\frac{3}{4} - 6\frac{1}{5}$ greater than or less than $13 - 6 = 7$?

87. _____

88. Is $1{,}629 - 976$ greater than or less than $1{,}600 - 1{,}000 = 600$?

88. _____

89. ★ ✎ Tara estimates $9\frac{1}{3} - 2\frac{3}{7}$ as $9 - 2 = 7$. Can you tell if 7 is an overestimate or an underestimate of $9\frac{1}{3} - 2\frac{3}{7}$ without making an additional computation? Explain.

PRACTICE | Fill each circle below with < or > to indicate which expression is greater without computing the actual differences.

90. $585 - 217 \bigcirc 600 - 200$

91. $1{,}219 - 572 \bigcirc 1{,}200 - 600$

92. $1{,}491 - 718 \bigcirc 1{,}500 - 700$

93. $300 \bigcirc 907 - 587$

94. $14 \bigcirc 41\frac{7}{13} - 28\frac{1}{8}$

95. $15\frac{2}{3} - 6\frac{2}{7} \bigcirc 10$

PRACTICE | Answer each question about overestimates and underestimates below.

96. When asked to estimate $813 - 189$, Yerg first adds 11 to 813 and to 189. Then, he computes $824 - 200 = 624$. Is Yerg's answer an **estimate**? Explain.

97. Is $947 - 658$ closer to 200 or to 300? Explain.

Use the following for the problems below:
Timmy estimates a difference by rounding. His estimate of the larger number is 7 more than the actual number. His estimate of the smaller number is 41 less than the actual number.

98. ★ When Timmy subtracts the rounded numbers, is his estimate more or less than the actual difference?

98. _____

99. ★★ When Timmy subtracts his rounded numbers, he gets 700. What is the actual difference between Timmy's numbers?

99. _____

When estimating a quotient, it helps to use numbers that are easy to divide!

EXAMPLE | Estimate $268 \div 7$.

When estimating a quotient, we look for numbers that make the division easier.

Since $40 \times 7 = 280$, we know that $280 \div 7 = 40$.
268 is close to 280, so $268 \div 7$ is close to $280 \div 7 = \mathbf{40}$.

The actual value of $268 \div 7$ is $38\frac{2}{7}$.
So, 40 is a good estimate.

PRACTICE | Use estimation to solve each problem below.

100. Estimate $816 \div 9$.

100. _____

101. Estimate $333 \div 8$.

101. _____

102. Estimate $34\frac{1}{9} \div 6\frac{7}{8}$.

102. _____

103. How many digits are in the quotient of $56{,}808 \div 789$?

103. _____

104. Circle the number below that is equal to $6{,}237 \div 11$ without computing the exact answer.

234   567   8,917   1,117

PRACTICE | Use estimation to solve each problem below.

105. Circle the number below that is equal to $12{,}321 \div 37$ without computing the exact answer.

111   333   999   5,555

106. Sean divides a 6-digit number by 20 and gets a 3-digit quotient. Explain why Sean's quotient cannot be correct.

107. Yerg estimates $768 \div 8$ by dividing $800 \div 8$.
Plunk estimates $768 \div 8$ by dividing $770 \div 10$.
Drew estimates $768 \div 8$ by dividing $770 \div 7$.
List the three estimates in order from least to greatest.

107. _____ _____ _____

108. Circle the two expressions below that have the same quotient.

$647 \div 82$ $6{,}478 \div 82$ $6{,}456 \div 8$ $64{,}859 \div 821$

109. Is it possible to divide a 4-digit number by a 2-digit number and get a 1-digit quotient? If so, give an example. If not, explain why it cannot be done.

EXAMPLE | Between which two consecutive multiples of 100 is the sum 268+765?

Since 200+700 = 900 gives us an underestimate, and 300+800 = 1,100 gives us an overestimate, we know that 268+765 is between 900 and 1,100.

We can improve our estimate by rounding both numbers to the nearest multiple of 50.
Since 268 is greater than 250 and 765 is greater than 750 we know that 268+765 is greater than 250+750 = 1,000.

So, 268+765 is **between 1,000 and 1,100.**

— *or* —

We can estimate by adding hundreds first: 200+700 = 900. Then, we estimate the remaining sum: 68+75 is more than 100, but less than 200. This means that 268+765 is between 900+100 and 900+200.

So, 268+765 is **between 1,000 and 1,100.**

> *Consecutive* means that one comes right after the other. For example, 5 and 6 are consecutive whole numbers, 40 and 45 are consecutive multiples of 5, and 1,000 and 1,100 are consecutive multiples of 100.

PRACTICE | Solve each estimation problem below.

---

110. Between which two consecutive multiples of 1,000 is the sum 11,956+3,582?

110. between _____ and _____

111. Between which two consecutive whole numbers is the sum $4\frac{1}{6} + 7\frac{3}{8}$?

111. between _____ and _____

112. Between which two consecutive multiples of 10 is the quotient 857÷9?

112. between _____ and _____

113. Between which two consecutive multiples of 10,000 is the product 439×57?

113. between _____ and _____

Beast Academy Practice 3D

PRACTICE | Use estimation to connect each expression to the point that marks its value on the number line above it.

114.

$$4\frac{7}{9}-2\frac{1}{5} \qquad 4\frac{7}{10}+2\frac{3}{5} \qquad 1\frac{5}{6}\times2\frac{7}{8}$$

115.

50 60 70 80 90 100 110

$$596\div8 \qquad 273-175 \qquad 372\div7$$

116.

200 250 300 350 400 450

$$1{,}356\div6 \qquad 18\times19 \qquad 216+226$$

117.

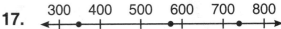

$$25\times23 \qquad 38\times9 \qquad 254+481$$

PRACTICE | For each problem below, use estimation to write the three expressions in order from least to greatest.

118. $168+326 \qquad 547-198 \qquad 19\times12$

118. _____ _____ _____

119. $3{,}116-2{,}338 \qquad 45\times9 \qquad 316+246$

119. _____ _____ _____

120. $414+318 \qquad 8{,}208\div36 \qquad 35\times19$

120. _____ _____ _____

In an Expression Maze, the goal is to draw a path that passes through every expression in order from least to greatest. Your path should begin at the arrow marked "Start" and end at the arrow marked "Finish."

You may only move up, down, left, or right to move between squares within the maze, and you may only visit each square once.

EXAMPLE | Complete the Expression Maze below.

```
        872              
        +345          ▶ Finish
                      
                   427
                   −193
Start▶      17       
          ×39        
                657  
                ÷81  
```

Since we only want to compare the expressions, we may not need to compute the exact values. We begin by estimating the value of each of the four expressions in the maze:

$872 + 345 \approx 900 + 300 = 1{,}200.$

$17 \times 39 \approx 20 \times 40 = 800.$

$657 \div 81 \approx 640 \div 80 = 8.$

$427 - 193 \approx 430 - 200 = 230.$

None of our estimates are close to each other, so we must visit the four expressions in this order:

Start ▶ | 657 ÷81 | ▶ | 427 −193 | ▶ | 17 ×39 | ▶ | 872 +345 | ▶ Finish

We can only accomplish this with the path shown below:

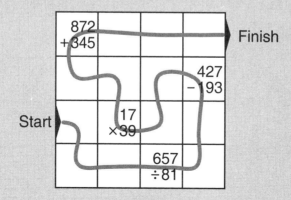

Beast Academy Practice 3D

PRACTICE | Complete each Expression Maze below.
We recommend you use a pencil.

121.

456 +465			645 +654
	546 +564		Finish ▶

Start

122.

31 ×27		13 ×39	
	11 ×23		
Start ▶			

Finish

123.

		258 −147	Finish ▶
	915 −575		
			659 −112

Start

124.

		145 ÷7	
	64 ÷5		Finish ▶
			59 ÷9
158 ÷10			

Start

125.

46 ×78			Finish ▶
	22 ×12		
		82 ×7	
53 ×18			

Start

126.

			996 +632
	682 +715	844 +319	
107 +712			
			Finish ▶

Start

PRACTICE | Complete each Expression Maze below.
We recommend you use a pencil.

127.

675 ×6			
		762 +915	
	172 ×3		
			837 +186

Finish

Start

128.

	50 ÷9	792 ÷18	
325 ÷40			175 ÷11

Start

Finish

129.

468 +326			
		226 ×3	
715 ÷6			
			33 ×11

Finish

Start

130.

Finish

	756 −192		453 −389
		8 ×19	
37 ×9			

Start

131. Do not pass through the gray square.

Finish

		12 ×27	
289 +225			
			579 ÷6
621 ÷72			

Start

132. Do not pass through the gray squares.

921 −389			
			635 +478
	758 ÷25		
		85 ×758	

Start

Finish

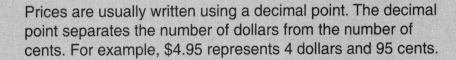

Prices are usually written using a decimal point. The decimal point separates the number of dollars from the number of cents. For example, $4.95 represents 4 dollars and 95 cents.

Estimating with money can be helpful for making quick calculations about costs!

When we estimate using money, we often round prices to the nearest dollar. Since there are 100 cents in a dollar, if the number of cents is less than 50, we round down. If the number of cents is 50 or more, we round up.

EXAMPLE | At the store, comics cost $3.95. Can Grogg buy five comics with twenty dollars?

Since $3.95 is just a little less than $4, five books will cost a little less than 5×4 = 20 dollars.

So, **yes, Grogg can buy five comics with twenty dollars**.

PRACTICE | Use the prices below for six books at the Beast Academy book fair to answer the questions that follow.

Where the Wild Things Are	$6.85	A Tale of Two Yetis	$1.95
Diary of a Wimpy Cyclops	$5.95	The Rancor in the Hat	$4.99
All the Pretty Jackalopes	$9.95	The Call of the Basilisk	$3.50

133. Can Lizzie buy three different books on this list for $11? If so, which three? If not, why?

134. Grogg buys two different books for a total cost of $9.45. Which two books did Grogg buy?

135. *Circle the correct ending to this sentence:*
The cost of all six books listed above is between:

$20 and $25 $25 and $30 $30 and $35 $35 and $40

136. Ms. Q. buys five copies of a book for $34.25. Which book did she buy five copies of?

136. _____

An educated guess that is based on how something looks is called an *eyeball estimate*.

EXAMPLE

Beast Island pennies are the same size as U.S. pennies, as shown below. About how many pennies could fit on one workbook page without overlapping?

If we lay the pennies in rows and columns to cover the page, we could then multiply the number of rows by the number of pennies in each row to figure out how many pennies it takes to cover the page. So, we guess how many pennies wide the page is, and how many pennies tall the page is, then multiply these two values.

Comparing the width of the penny to the width of the page, we guess that the page is about 10 pennies wide, and 15 pennies tall. So, we guess that it will take about $10 \times 15 = $ **150 pennies** to cover the page.

PRACTICE | Answer each eyeball estimation question below.

137. The square on the left can be painted with one ounce of paint. How many ounces of paint are needed to paint the shape on the right?

137. _____

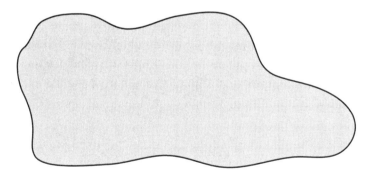

138. About how many dots are printed below?

138. _____

PRACTICE | Use the map below for the two questions that follow.

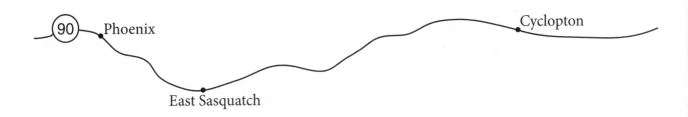

139. Along Highway 90, Phoenix is 17 miles from East Sasquatch. About how many miles is the distance from Phoenix to Cyclopton along Highway 90?

139. _____

140. Driving along Highway 90, it takes about 60 minutes to get from East Sasquatch to Cyclopton. About how many minutes will it take to get from Phoenix to East Sasquatch?

140. _____

141. The minute hand has fallen off of this clock! Using only the hour hand, estimate the time.

141. _____

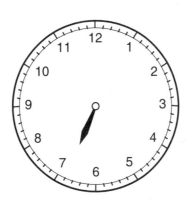

PRACTICE | Answer each eyeball estimation question below.

142. Grogg is almost 5 feet tall. About how tall is the Beast Island Lighthouse?

142. _____

143. The scalene right triangle below has a perimeter of 12 cm. Estimate the length of its shortest side without measuring.

143. _____

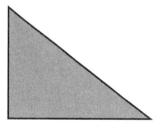

144. A quarter on Beast Island is almost exactly one inch across, as shown below. Estimate the length in inches of the line below without measuring.

144. _____

1 inch

PRACTICE | Answer each eyeball estimation question below.

145. A nickel is worth 5 cents. About how many dollars worth of nickels are shown below?

145. _____

PRACTICE | Beastberries cost $2.95 per pound. Estimate the cost in dollars of each bag of beastberries below.

146.

147.

146. _____

147. _____

Enrico Fermi was a Nobel-prize-winning physicist who enjoyed thinking up seemingly impossible questions and estimating their answers. Problems like these are often called *Fermi problems*.

To solve a Fermi problem, we make a few reasonable guesses in order to calculate an estimate.

EXAMPLE | How many slices of pizza were eaten in the United States yesterday?

No one can possibly know the exact answer, but we can make a rough estimate. It is useful to know that the population of the United States is a little over 300 million.

We consider two easier questions:
- About how many people in the United States ate pizza yesterday?
- About how many slices did each of those people eat?

Since 300 million is a big number, we consider a smaller group of 30 people. We'll guess that about 2 out of every 30 people ate pizza yesterday. Ten groups of 30 make a group of 300 people. So, in a group of 300 people, about $2 \times 10 = 20$ ate pizza yesterday. It takes a million groups of 300 people to make 300 million people. So, in a country of 300 million people, about 20 million people ate pizza yesterday.

Each person who ate pizza probably had about 2 or 3 slices. We can multiply 2 slices by 20 million to get 40 million slices, or 3 slices by 20 million to get 60 million slices of pizza eaten yesterday.

So, we estimate that **between 40 and 60 million slices of pizza** were eaten yesterday.

— *or* —

We can start with a different question:
- How many slices of pizza does one person eat in a week?

Some people eat pizza all the time, and others eat no pizza at all. The people who eat pizza might eat lots of pizza one week and no pizza another week. We guess that a typical person eats about 1 slice of pizza each week.

The population of the United States is a little over 300 million. If each person eats one slice of pizza each week, then about 300 million slices of pizza are eaten each week. Since there are 7 days in a week, we divide 300 million by 7 to figure out how many slices of pizza are eaten every day. $300,000,000 \div 7 \approx 280,000,000 \div 7 = 40,000,000$. So, we guess that **about 40 million slices of pizza** were eaten yesterday.

This is a pretty rough guess, but we are pretty sure that the answer isn't 1 thousand or 3 trillion. Any estimate between 10 and 100 million slices is reasonable.

PRACTICE | Make your best guess for each question below.

148. How many apples will fill a bathtub?

149. What is the length in feet that one of your fingernails will grow in your lifetime?

150. How many breaths will you take today?

151. How many commercials are played on a television channel each day from 6 a.m. to 10 p.m.?

152. If you could walk without stopping all day and all night, about how many days would it take for you to walk from Los Angeles to New York?
It may help to know that it is about 2,800 miles from Los Angeles to New York.

153. How many words are in this book?

CHAPTER 12
Area

Use this Practice book with
Guide 3D from BeastAcademy.com.

Recommended Sequence:

Book	Pages:
Guide:	74 – 85
Practice:	77 – 84
Guide:	86 – 109
Practice:	85 – 103

You may also read the entire chapter
in the Guide before beginning the
Practice chapter.

Area is the amount of space a shape takes up on the plane.

Area is given in **square units**, like square centimeters.

A **square centimeter** is the area of a square with sides that are 1 centimeter long. "Square centimeter" can be written **sq cm** for short.

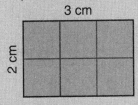 = 1 square centimeter (sq cm)

EXAMPLE | What is the area of the rectangle below?

3 cm
2 cm

The rectangle has an area of 2×3 = **6 square centimeters (6 sq cm)**.

PRACTICE | Each rectangle below is drawn actual size. Write the area of each rectangle in square centimeters (sq cm).

1.
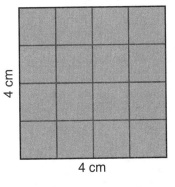
4 cm
4 cm

2.
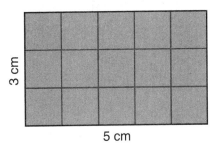
3 cm
5 cm

3.
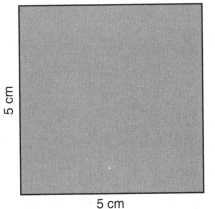
5 cm
5 cm

4.
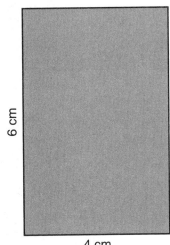
6 cm
4 cm

1. _____

2. _____

3. _____

4. _____

Units of Area
Abbreviations are in parentheses.

Square Inches (sq in)
Square Feet (sq ft)
Square Yards (sq yd)
Square Miles (sq mi)
Square Centimeters (sq cm)
Square Meters (sq m)
Square Kilometers (sq km)

Each unit of length has a corresponding unit of area.

For example, a square that has one-inch sides has an area of one square inch (sq in).

Here are some common units of area and their abbreviations.

When lengths are given without units, area is given in square units (sq units).

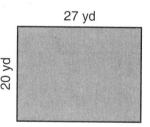

PRACTICE | Write the area of each rectangle below using the correct units. Rectangles are not actual size.

5.
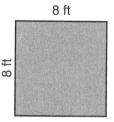
8 ft
8 ft

5. _____

6.
15 m
10 m

6. _____

7.
27 yd
20 yd

7. _____

8.
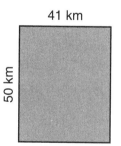
41 km
50 km

8. _____

9.
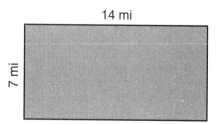
14 mi
7 mi

9. _____

10.
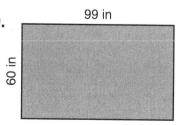
99 in
60 in

10. _____

PRACTICE | Answer the following questions. Be sure to write the correct units (or square units) with your answer.

11. What is the area of a square that has 9-centimeter sides?

11. _____

12. The rectangular floor of Ms. Q.'s classroom is 40 feet long and 30 feet wide. What is the area of Ms. Q.'s classroom floor?

12. _____

13. The tabletops in Captain Kraken's woodshop are 60 inches long and 30 inches wide. What is the area of one of Captain Kraken's tabletops?

13. _____

14. At Beast Academy, the rectangular Beastball field is 30 yards wide and 100 yards long. What is the area of the Beastball field?

14. _____

15. What is the **_perimeter_** of a rectangle that is 6 inches wide and has an area of 42 square inches?

15. _____

16. ★ A square is split into three congruent rectangles, as shown. The perimeter of each rectangle is 16 inches. What is the area of the square?

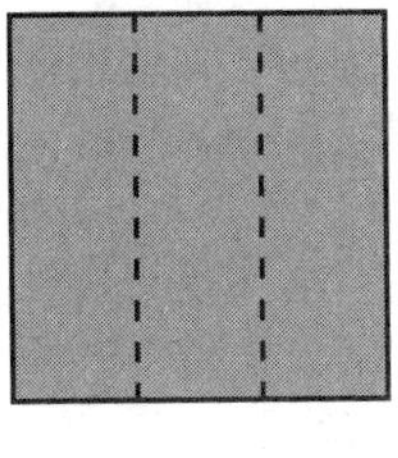

16. _____

To find the area of a shape, you can split the shape into parts whose areas are easy to compute.

The area of the shape is the sum of the areas of its parts.

EXAMPLE

Two squares are attached to make the shape below. What is the area of the shape they form?

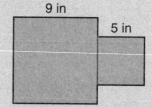

9 in

5 in

The square on the left has an area of 9×9 = 81 sq in. The square on the right has an area of 5×5 = 25 sq in.

The total area of the shape is 81+25 = **106 sq in**.

PRACTICE | Solve each area problem below.

---

17. Three squares are arranged as shown below to make a rectangle. The area of the shaded square is 9 square miles. What is the area of the rectangle?

17. _____

---

18. The shape below was made by joining 9 congruent squares. What is the area of the shape?

18. _____

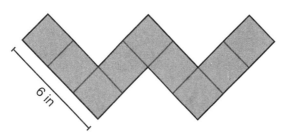

6 in

---

19. The letter "T" below was made by joining two congruent rectangles as shown. What is the area of the "T"?

19. _____

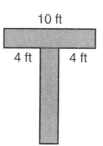

10 ft

4 ft   4 ft

Remember, a shape is **rectilinear** if its sides always meet at right angles.

PRACTICE | Each rectilinear shape below can be split into two or more rectangles. Find the area of each shape.

20.

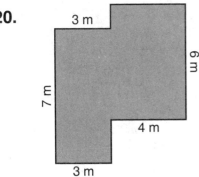

3 m
6 m
7 m
4 m
3 m

20. _____

21.

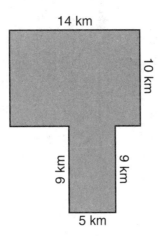

14 km
10 km
9 km
9 km
5 km

21. _____

22.

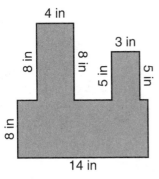

4 in
8 in
8 in
3 in
5 in
5 in
8 in
14 in

22. _____

23.

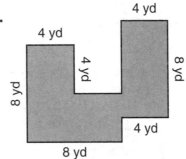

4 yd
4 yd
4 yd
8 yd
8 yd
4 yd
8 yd

23. _____

24. Four congruent rectangles are connected to form a large rectangle as shown below. What is the area of the large rectangle?

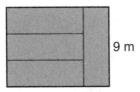

9 m

24. _____

EXAMPLE | Find the area of the rectilinear shape below.

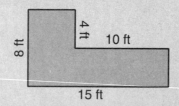

Sometimes it is easiest to use subtraction to compute the area of a shape.

We think of the figure as a large rectangle with a smaller rectangle removed, as shown.

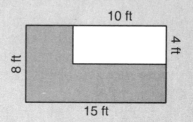

The large rectangle has an area of $8 \times 15 = 120$ square feet. The small rectangle has an area of $4 \times 10 = 40$ square feet. The area of the shape formed by removing the small rectangle from the large one is $120 - 40 = \textbf{80 square feet}$.

We could also split the shape into two rectangles and add their areas to get the same total area.

PRACTICE | Solve each area problem below.

25. The 8 inch by 10 inch rectangle below has a square hole with 4-inch sides cut out of it. What is the area of the remaining shape?

25. _____

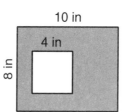

26. Find the area of the rectilinear shape below.

26. _____

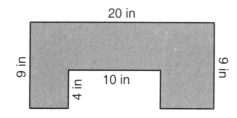

PRACTICE | Find the area of each rectilinear shape below.

27.

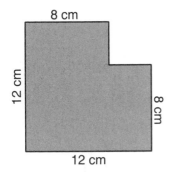

28.

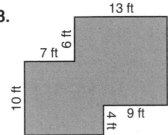

27. _____

28. _____

29. Congruent squares are arranged to create the shape below. Each square has a side length of 2 cm. What is the area of the entire shape?

29. _____

30. A square is split into four congruent triangles. One triangle is removed, leaving the shape below. What is the area of the remaining shape?

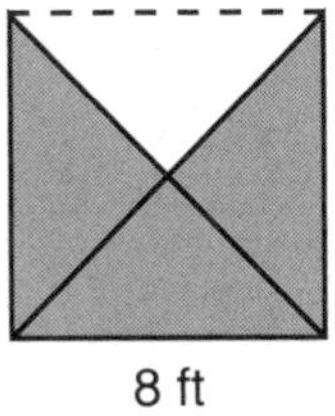

30. _____

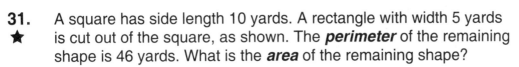

8 ft

31.
★ A square has side length 10 yards. A rectangle with width 5 yards is cut out of the square, as shown. The **perimeter** of the remaining shape is 46 yards. What is the **area** of the remaining shape?

31. _____

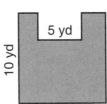

5 yd

10 yd

PRACTICE | Solve each area problem below.

32. Four congruent quadrilaterals are connected as shown to form a 10 cm by 10 cm square with a hole in the middle. Each quadrilateral has an area of 17 sq cm. What is the area of the square hole?

32. _____

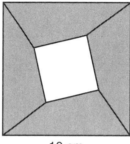

10 cm

33. Four squares are connected to form the rectangle below with a rectangular hole in the middle. Find the area of the shaded region.

33. _____

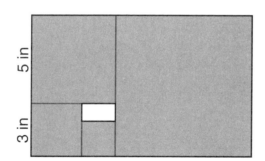

5 in

3 in

34. ★ The three squares in the figure below have side lengths of 3 cm, 5 cm, and 7 cm. Find the total shaded area.

34. _____

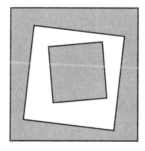

The **tangram** is an ancient Chinese puzzle in which a square is split into 7 pieces, called tans, as shown below. The seven tans can be rearranged to form thousands of new shapes, each with the same area as the original square.

The challenge is to arrange the 7 tans to match the outline of a given shape. All 7 tans must be used, and tans may not overlap. For example, the seven tans can be rearranged as shown below to create a house, a sailboat, or a swan.

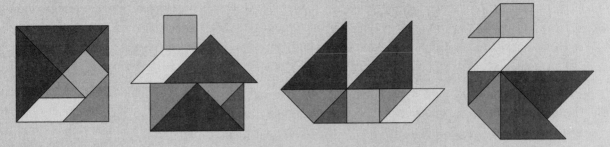

Carefully cut along the lines below to create your own set of tans. Arrange the 7 tans to create each of the samples above. Then, try to create each of the shapes on the next page.

Guide Pages: 86-87

PRACTICE | Solve each tangram puzzle below. Be sure to use *all* 7 tans for each figure!

35. Bunny

36. Zombie

37. Meerkat

38. Fish

39. Triangle

40. Rectangle

41. ★ Square with a missing triangle

42. ★ Quadrilateral with a missing quadrilateral

Any rectangle can be split into two right triangles.

Two congruent right triangles can always be arranged to make a rectangle.

To find the area of any right triangle, we think of the triangle as half of a rectangle.

EXAMPLE | Find the area of the right triangle below.

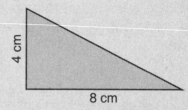

4 cm

8 cm

We can arrange two copies of the triangle above to make a rectangle as shown.

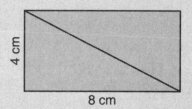

4 cm

8 cm

The rectangle has an area of $4 \times 8 = 32$ sq cm. The triangle is half the area of the rectangle. So, the area of the triangle is $32 \div 2 =$ **16 sq cm**.

PRACTICE | Find the area of each right triangle below.

43.

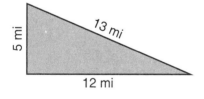

5 mi
13 mi
12 mi

44.

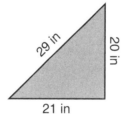

29 in
20 in
21 in

43. _____

44. _____

45.

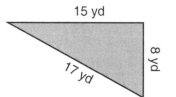

15 yd
8 yd
17 yd

46.

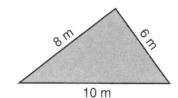

8 m
6 m
10 m

45. _____

46. _____

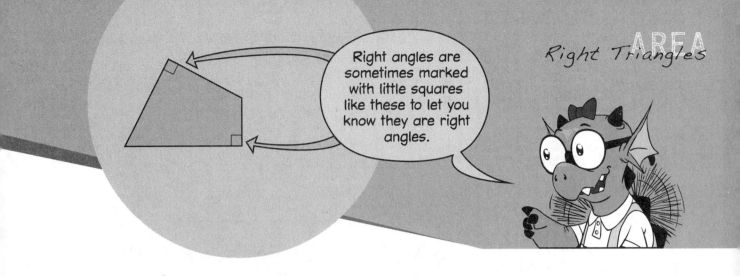

PRACTICE | Find the area of each shape below.

47.

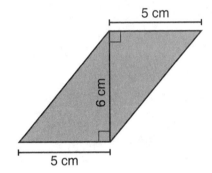

48.

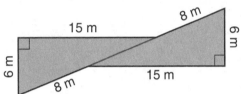

47. _____

48. _____

49.

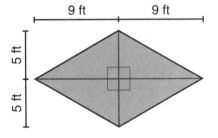

50.

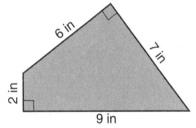

49. _____

50. _____

51.
★

Two right triangles overlap to make the shape below. The area where the triangles overlap is a 4 inch by 4 inch square. What is the area of the whole shape?

51. _____

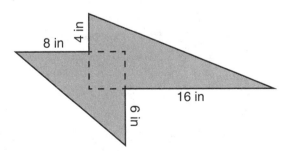

In each dot grid below, each dot is 1 unit from its nearest horizontal and vertical neighbor.

We can use right triangles to find the area of a shape drawn on a dot grid.

EXAMPLE Find the area of the quadrilateral traced on the dot grid below.

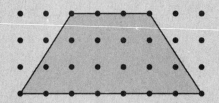

We can split the shape into three parts: two triangles and one square.

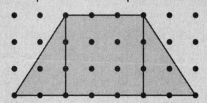

Each dot is one unit from its nearest horizontal and vertical neighbor.

The square has an area of $3\times3=9$ square units.
Each triangle is half of a 2 by 3 rectangle. So, the combined area of the triangles is $2\times3=6$ square units.

All together, the shape has an area of
$9+6=$ **15 square units**.

PRACTICE | Find the area of the shaded region on each dot grid below.

52.

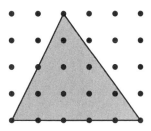

53.

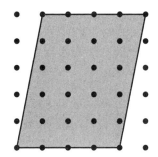

52. _____

53. _____

54.

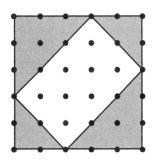

55.

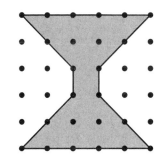

54. _____

55. _____

PRACTICE | Find the area of the shaded region on each dot grid below.

56.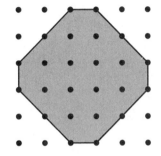

57.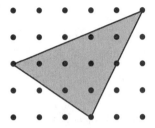

56. _____

57. _____

58.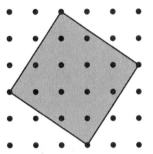

59.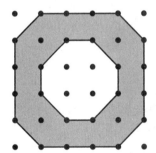

58. _____

59. _____

60.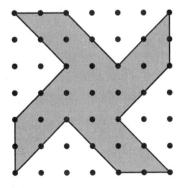

61. ★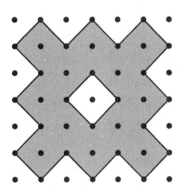

60. _____

61. _____

Connectahalf is a pencil-and-paper game for two players.

The game is played on an enclosed dot grid. You may use one of the grids on the facing page, make your own, or print more grids at BeastAcademy.com.

In Connectahalf, players take turns connecting points on a dot grid to create a "cut line" that divides the shape into two pieces. The goal is to be the player who completes the cut line so that it splits the shape into two parts of equal area.

Sample Game: The sample game below describes the rules of play.

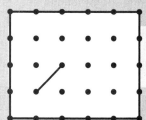

Player 1 connects any two neighboring points on the dot grid (horizontally, vertically, or diagonally). This begins the "cut line."

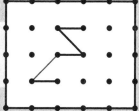

Players take turns extending the cut line by connecting one end to a neighboring point. After 5 turns, the sample game looks like this.

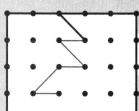

Next, Player 2 connects the cut line to the top side of the rectangle.

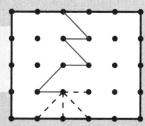

Player 1 now has four choices. Three of the choices complete the cut line, but none split the rectangle into two parts of equal area.

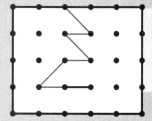

If a player completes a cut line that does **not** split the shape into two equal areas, the other player wins the game. Player 1 extends the cut line to avoid losing.

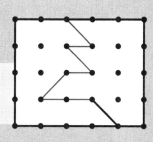

Player 2 connects the cut line to the bottom of the rectangle as shown. This completes the cut line.

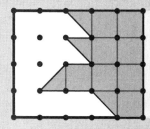

Players compute the area on either side of the cut line. To make it easier, the area can be split into whole and half squares as shown.

The shaded area has 8 whole squares and 4 half squares.
The 4 half squares can be arranged to make 2 whole squares.
So, the shaded area is 8 + 2 = 10 square units.

The whole rectangle has an area of 4 × 5 = 20 square units, so the unshaded area is also 20 − 10 = 10 square units. Since the cut line divides the rectangle into two parts of equal area, Player 2 wins!

PRACTICE | Find the winning move on each Connectahalf grid below.

62.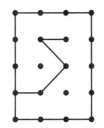

63.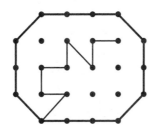

64.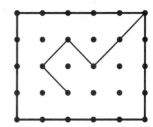

PRACTICE | Find a partner and play Connectahalf on each of the game boards below.

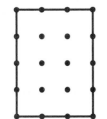
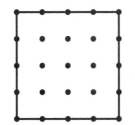
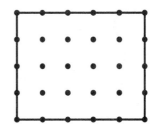

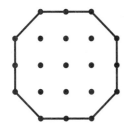
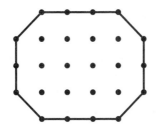
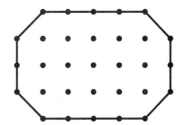

PRACTICE | Solitaire Connectahalf: Split each of the shapes below into two parts of equal area using the fewest moves possible.

65.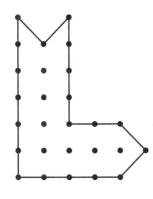

66.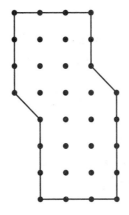

67.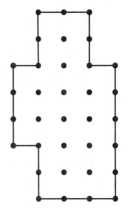

AREA
Triangles

EXAMPLE | Find the area of the triangle below.

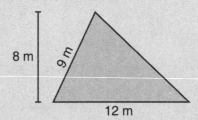

To find the area of any triangle, multiply the length of its base by its height, then divide by 2.

Sometimes, it is useful to turn the triangle so that the base is on the bottom.

The **base** is the side that we measure the height from. The base of the triangle above is 12 meters long.

The **height** of the triangle is measured straight up from the base. The height of this triangle is 8 meters.

The area of the triangle is found by:
Area = base × height ÷ 2

So, the area of the triangle above is
12×8÷2 = 96÷2 = **48 sq m**.

PRACTICE | Find the area of each triangle below.

68.

10 ft, 12 ft, 8 ft

69.

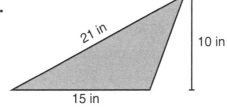

21 in, 10 in, 15 in

68. _____

69. _____

70.

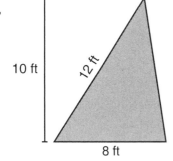

20 cm, 12 cm, 13 cm, 21 cm

71. ★

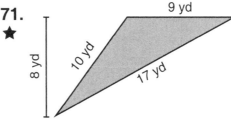

9 yd, 8 yd, 10 yd, 17 yd

70. _____

71. _____

PRACTICE | Solve each area problem below.

72. What is the height of a triangle that has an area of 45 sq ft and a base length of 15 ft?

72.

73. Use a centimeter ruler to find the area of the triangle below in square centimeters.

73.

74. ★ Five congruent triangles are joined to make the quadrilateral below. What is the area of the quadrilateral?

74.

8 in

21 in

75. ★ The area of triangle ABC below is 104 sq cm. What is the area of triangle ABD?

75.

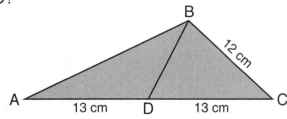

B

12 cm

A   13 cm   D   13 cm   C

In each Fragment Puzzle below, the goal is to find the pieces that can fit together to create a given shape. Pieces may be flipped or rotated.

EXAMPLE | Which three pieces below can be arranged to make a 3 unit by 3 unit square?

You can print larger versions of these puzzles to cut out at BeastAcademy.com.

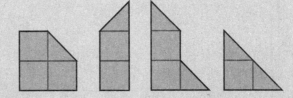

The area of a 3 unit by 3 unit square is 9 square units, so the three pieces used to make the square must have a combined area of 9 square units.

In the four pieces above, there are a total of 8 whole squares and 6 half squares. We can combine the 6 half squares to make 3 whole squares. So, the total area of all four pieces is 8+3 = 11 square units. That is two square units more than we need!

If we remove the piece that has an area of 2 square units, we will be left with three pieces whose total area is 11−2 = 9 square units.

Only the piece on the right (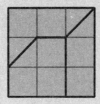) has an area of 2 square units: 1 whole square + 2 half squares.

The other three shapes can be arranged to create a 3 unit by 3 unit square, as shown below.

PRACTICE | Circle the three pieces below that can be arranged to make a 3 unit by 4 unit rectangle.

76.

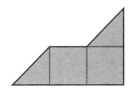
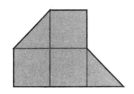
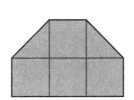
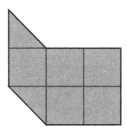

Beast Academy Practice 3D

PRACTICE In each fragment puzzle below, a target shape is given on the left. Circle the three pieces on the right that can be arranged to make the target shape.

For larger printable versions of these puzzles, visit BeastAcademy.com.

77.

78.

79.

80.

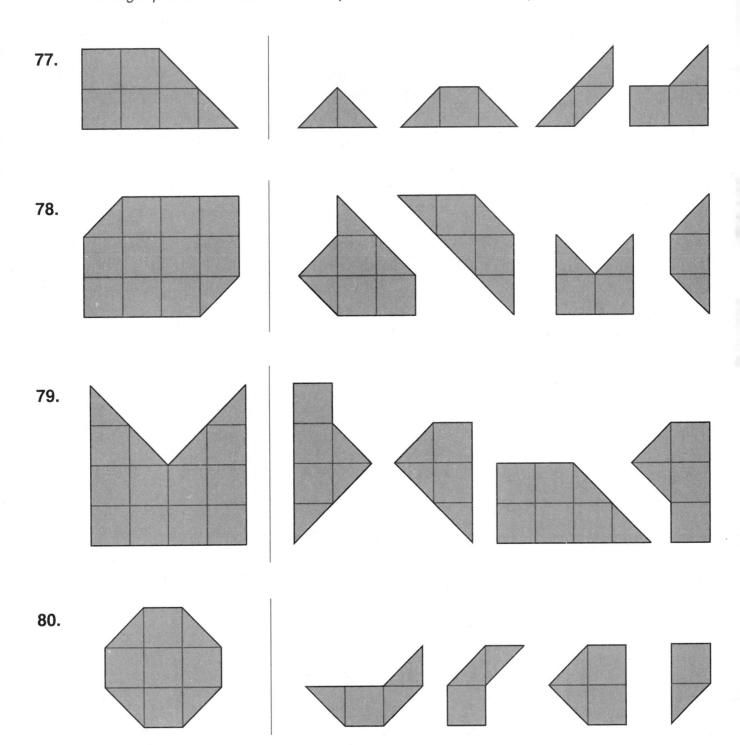

AREA

Different Units

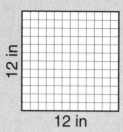

EXAMPLE | How many square inches are there in one square foot?

A square with an area of 1 square foot has sides that are 1 foot long. Since there are 12 inches in a foot, the sides of the square are 12 inches long.

12 in

12 in

So, there are 12×12 = **144 square inches** in 1 square foot.

PRACTICE | Solve each area problem below.

81. How many square centimeters are there in one square meter?

81. _____

82. Circle the number below that is closest to the number of square feet in a square mile. (1 mile = 5,280 feet)

5,000   25,000   5,000,000   25,000,000   250,000,000

83. The number of square inches in one square yard is between

0 and 100   100 and 400   400 and 900   900 and 1,600

84. How many square feet are there in 4 square yards?

84. _____

85. A rectangular sidewalk is one yard wide and one mile long. How many square feet are in the area of the sidewalk?

85. _____

98

PRACTICE | Solve each area problem below.

86. Rosencrantz and Guildenstern want to lay 4 inch by 4 inch square
★ tiles to completely cover an 8 foot by 10 foot bathroom floor. How
many tiles will they need to cover the floor?

86. _____

87. Captain Kraken needs to paint an area of 15 square yards on his
★ ship. One ounce of paint will cover 3 square *feet*. How many ounces
of paint will it take to cover 15 square yards?

87. _____

88. How many 3 inch by 5 inch index cards will it take to completely
★ cover the surface of a table that is 3 feet wide and 5 feet long?

88. _____

89. Ms. Levans needs 50 square feet of fabric to make a quilt. The Beast
★ Island fabric depot sells fabric by the square yard, and Ms. Levans
can only buy a whole number of square yards of fabric. How many
square yards of fabric must Ms. Levans buy to be sure she has at
least 50 square feet of fabric?

89. _____

90. Cammie has a rectangular sheet of construction paper that is
★ 8 inches tall and has an area of 1 square foot. How many inches
wide is Cammie's sheet of paper?

90. _____

PRACTICE Solve each area problem below.

91. What is the area of the quadrilateral below?

91. _____

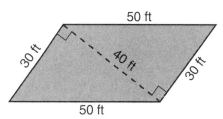

92. The square and triangle below have the same area. What is the height of the triangle, measured from the base as shown?

92. _____

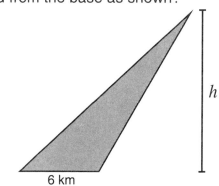

93. The side length of square EFGH is twice the side length of square ABCD. The area of square ABCD is 7 sq m. What is the area of square EFGH?

93. _____

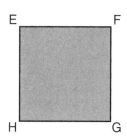

94. What is the area in **square yards** of a rug that is 10 feet wide and 18 feet long?

94. _____

PRACTICE

Square PQRS below is divided into 8 congruent triangles. The area of square PQRS is 28 sq ft.

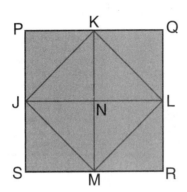

95. What is the area of square PKNJ?

95. _____

96. What is the area of square JKLM?

96. _____

PRACTICE

Four congruent right triangles with the dimensions given on the left are arranged around a small square to make a large square as shown below.

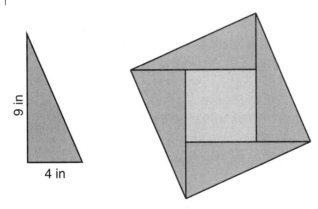

9 in

4 in

97. ★ What is the area of the small square?

97. _____

98. ★ What is the area of the large square?

98. _____

99. When the 4-inch side of the triangle below is used as its base, the height of the triangle is 15 inches. What is the height of the same triangle when the 20-inch side is used as its base?

99. _____

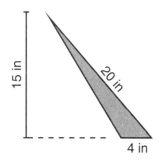

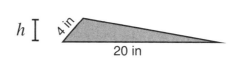

100. Circle the four triangles below that can be arranged to make an 8 foot by 8 foot square.

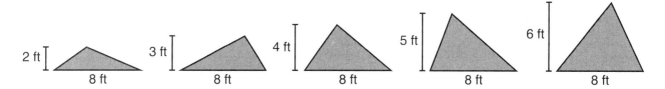

101. A triangle and square are attached to create a pentagon with height 24 meters, as shown. The triangle has the same area as the square. What is the side length of the square?

101. _____

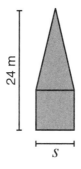

102. Four congruent right triangles are arranged as shown below to make a 10 meter by 10 meter square with a square hole in the middle. What is the area of the square hole?

★

102. _____

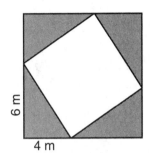

103. Each dot in the grid below is 1 unit away from its nearest horizontal and vertical neighbor. Draw a line from the top of the triangle below to a point on its bottom side so that the line splits the triangle into two smaller triangles of equal area.

★

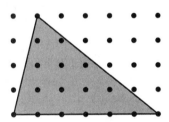

104. The area of the square below is 16 square inches. The square is split into 7 tans, each labeled with a letter. The tan labeled ⓓ is a square. Find the area of each tan.

★

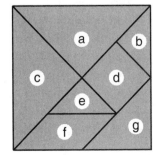

a. _____   b. _____   c. _____

d. _____   e. _____   f. _____

g. _____

HINTS
For Selected Problems

Below are hints to every problem marked with a ★.
Work on the problems for a while before looking at the hints.
The hint numbers match the problem numbers.

CHAPTER 10
Fractions   6–45

33. Start by connecting 7 to an equal number with a straight line. How can you connect the other pairs of equal numbers around that?

61. There are six possible ways to arrange the digits 1, 2, and 3 to create a fraction that is greater than 1. Organize these fractions by denominator, then evaluate.

62. There are six possible ways to arrange the digits 4, 5, and 6 to create a fraction that is greater than 1. Organize these fractions by denominator, then evaluate.

63. Which two whole numbers is $\frac{20}{3}$ between? Which two whole numbers is $\frac{40}{3}$ between?

66. Begin by locating the whole numbers on the number line.

72. Start by labeling the whole numbers on the number line.

125. All of the fractions are equal to $\frac{2}{5}$.

127. $\frac{1}{s} \xlongequal{\times?} \frac{s}{16}$ (with $\times?$)

128. We do not need to know the values of a and b to solve this problem. What number must we multiply a by to get b?

155. Begin by simplifying all fractions.

156. Begin by simplifying all fractions.

157. Begin by simplifying all fractions.

158. Begin by simplifying all fractions.

162. Where could we put point D to make it as far as possible from point A?

163. Where could we put point D to make it as close as possible to point A?

167. How could you draw and measure a line that is twice as long as the distance from Bobby's house to Carl's house?

168. The total distance from Bobby's house to Carl's house is $4\frac{1}{4} = 4\frac{2}{8}$ inches.

193. Where can each number be placed so that each is the denominator of a fraction in simplest form?

194. Which numerator makes the fraction with denominator 7 less than 1?

195. Which numbers can be placed below the 2 in the middle fraction?

199. We can convert these two fractions to equivalent fractions with equal numerators (or denominators) to compare them. For example,

$$\frac{2}{5} \xlongequal{\times 7} \frac{14}{35} \quad \text{and} \quad \frac{7}{21} \xlongequal{\times 2} \frac{14}{42}$$

or

$$\frac{2}{5} \xlongequal{\times 21} \frac{42}{105} \quad \text{and} \quad \frac{7}{21} \xlongequal{\times 5} \frac{35}{105}$$

200. $\frac{37}{99} \xlongequal{\times 2} \frac{74}{198}$

Can you tell which is bigger, $\frac{74}{198}$ or $\frac{73}{200}$?

201. First, convert both fractions so that their numerators are the same:

$$\frac{2}{7} \xlongequal{\times 3} \frac{6}{21} \quad \text{and} \quad \frac{3}{8} \xlongequal{\times 2} \frac{6}{16}$$

What unit fraction is between $\frac{6}{21}$ and $\frac{6}{16}$?

202. There are six possible ways to arrange the digits 7, 8, and 9 to create a fraction that is less than 1. Organize these fractions by numerator, then compare each to $\frac{1}{10}$ and $\frac{1}{9}$.

210. This shape can be split into 4 congruent pentagons.

211. This shape can be split into 3 congruent quadrilaterals.

212. This shape can be split into 7 congruent right triangles.

CHAPTER 11
Estimation   46–75

27. Which numbers round to 8,000 when rounded to the nearest ten? When rounded to the nearest hundred? When rounded to the nearest thousand?

28. Which numbers round to 400 when rounded to the nearest hundred? Which numbers round to 450 when rounded to the nearest ten?

29. Work backwards. What is the smallest possible value of Lizzie's number?

66. Begin by connecting 168×38 to the equal expression on the right with a straight line. How can you connect the other pairs of expressions around this wire?

67. Begin by connecting 97×34 to the equal expression on the right with a straight line. How can you connect the other pairs of equal expressions around this wire?

68. Begin by connecting 13×87 to the equal expression on the right with a straight line. How can you connect the other pairs of equal expressions around this wire?

69. Begin by connecting 91×19 to the equal expression on the right with a straight line. How can you connect the other pairs of equal expressions around this wire?

70. Rewrite each fraction as a mixed number and begin by estimating the value of each sum.

Then, when pairing numbers, start by connecting $\frac{41}{4}+\frac{37}{8}$ to its equal expression on the right with a straight line. How can you connect the other pairs of expressions around this wire?

71. Begin by estimating the value of each product.

Then, when pairing numbers, start by connecting 78×77 directly to the equal expression on the right with a straight line. How can you connect the other pairs of expressions around this wire?

76. Try some examples. If Norton rounded to the nearest ten before multiplying, would his estimate of 9×13 be an overestimate or underestimate? How about 7×14?

84. What is 35×20?

85. Will Rosencrantz and Guildenstern need more or less than 100 blocks?

89. Can you tell whether the rounded values are closer together or farther apart than the original values?

98. Are the numbers that Timmy subtracts closer together or farther apart on the number line than the original numbers?

99. *By how much* are Timmy's rounded numbers closer together or farther apart than the original numbers?

148. An apple is about 3 inches long, 3 inches tall, and 3 inches wide. What are the dimensions, in inches, of a bathtub?

149. How long does a fingernail grow in one year?

150. How many breaths do you take in one minute?

151. About how many minutes of commercials are shown each hour? About how long is each commercial?

152. A normal walking speed is about 2 miles per hour.

153. Some pages in this book have lots of words, but others have very few. The solutions pages have many more words than the problems pages, so you might estimate solution pages and problem pages separately.

16. What is the relationship between the height and width of each congruent rectangle?

31. What is the total length of the three horizontal sides on top of the shape?

34. What is the area of the outer shaded region, not including the small shaded square?

41. One of the large triangles should be placed in the lower-right corner, as shown:

42. One of the large triangles should be placed in the lower-left corner, as shown:

51. Begin by finding the area of each triangle.

61. We can split the shape into twelve congruent squares, as shown:

What is the area of each square?

71. Which side is the height of the triangle measured from?

74. What is the area of one of the five congruent triangles?

75. How does the base length and height of triangle ABD compare to the base length and height of triangle BCD?

86. How many 4 inch by 4 inch square tiles will cover *one* square foot? Make a sketch!

87. How many ounces of paint are needed to paint *one* square yard?

88. How many cards can be placed along the short side of the table? The long side? Make a sketch!

89. Since Ms. Levans can only buy a whole number of square yards, how many square feet can she buy? For example, can she buy exactly 7 square feet of fabric?

90. What is the area of Cammie's paper in square *inches*?

97. How could you find the side length of the small square?

98. How could you find the area of the square without knowing its side length?

99. We can use *any* side of a triangle as the base. We can use the length of that base and the height measured from that base to calculate the triangle's area.

100. What is the area of an 8 ft by 8 ft square?

101. How does the height of the triangle compare to the height of the square?

102. What is the total area of the four shaded triangles?

103. Two triangles created by drawing a line from the top of the triangle below to a point on the bottom side will have the same height (when measured from the bottom side).

104. Place the original square on a 4 inch by 4 inch dot grid.

SOLUTIONS
Chapters 10-12

FRACTIONS
Fractions are Numbers!   *page 7*

1. The number line between 0 and 1 has been split into four equal pieces, so each piece has length $\frac{1}{4}$. The first piece begins at 0 and ends at $\frac{1}{4}$.

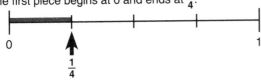

2. The number line between 0 and 1 has been split into seven equal pieces, so each piece has length $\frac{1}{7}$. The first piece begins at 0 and ends at $\frac{1}{7}$.

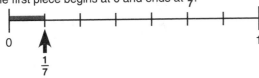

3. The number line between 0 and 1 has been split into eight equal pieces, so each piece has length $\frac{1}{8}$. The first piece begins at 0 and ends at $\frac{1}{8}$.

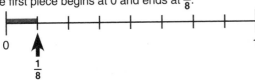

4. The number line between 0 and 1 has been split into ten equal pieces, so each piece has length $\frac{1}{10}$. The first piece begins at 0 and ends at $\frac{1}{10}$.

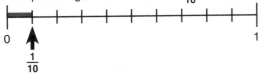

FRACTIONS
Comparing Unit Fractions   8

5. We use the marked number lines to locate $\frac{1}{2}$ and $\frac{1}{3}$:

Since $\frac{1}{2}$ is to the right of $\frac{1}{3}$ on the number line, $\frac{1}{2}$ is greater than $\frac{1}{3}$.

6. We use the marked number lines to locate $\frac{1}{6}$ and $\frac{1}{9}$:

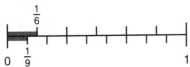

Since $\frac{1}{6}$ is to the right of $\frac{1}{9}$ on the number line, $\frac{1}{6}$ is greater than $\frac{1}{9}$.

7. We use the marked number lines to locate $\frac{1}{10}$ and $\frac{1}{11}$:

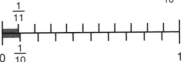

Since $\frac{1}{10}$ is to the right of $\frac{1}{11}$ on the number line, $\frac{1}{10}$ is greater than $\frac{1}{11}$.

8. We use the marked number lines to locate $\frac{1}{40}$ and $\frac{1}{29}$:

Since $\frac{1}{29}$ is to the right of $\frac{1}{40}$ on the number line, $\frac{1}{29}$ is greater than $\frac{1}{40}$.

9. We can compare $\frac{1}{91}$ and $\frac{1}{100}$ without a number line. Looking at the problems above, we see that the more pieces we divide the line between 0 and 1 into, the smaller each piece must be. The fraction $\frac{1}{91}$ is the length of each piece when the number line between 0 and 1 is split into 91 equal pieces. The fraction $\frac{1}{100}$ is the length of each piece when the number line between 0 and 1 is split into 100 equal pieces. Since 100 pieces is more than 91 pieces, the pieces of length $\frac{1}{100}$ are smaller than the pieces of length $\frac{1}{91}$. So, $\frac{1}{91}$ is greater than $\frac{1}{100}$.

*Notice that in all of the above problems, **the unit fraction with the smaller denominator is greater**. This is because the more pieces we divide the same part of the number line into, the smaller each piece must be.*

FRACTIONS
Whole-Number Fractions   9

10. $\frac{16}{8} = 16 \div 8 = \mathbf{2}$.

11. $\frac{48}{4} = 48 \div 4 = \mathbf{12}$.

12. $\frac{45}{9} = 45 \div 9 = \mathbf{5}$.

13. $\frac{75}{25} = 75 \div 25 = \mathbf{3}$.

14. $\frac{39}{13} = 39 \div 13 = \mathbf{3}$.

15. $\frac{12}{12} = 12 \div 12 = \mathbf{1}$.

16. The missing numerator is the number that can be divided by 3 to get 2. Since $\boxed{6} \div 3 = 2$, we have $\frac{\mathbf{6}}{3} = 2$.

17. The missing numerator is the number that can be divided by 7 to get 10. Since $\boxed{70} \div 7 = 10$, we have $\frac{\mathbf{70}}{7} = 10$.

18. The missing numerator is the number that can be divided by 3 to get 12. Since $\boxed{36} \div 3 = 12$, we have $\frac{\mathbf{36}}{3} = 12$.

19. The missing numerator is the number that can be divided by 6 to get 9. Since $\boxed{54} \div 6 = 9$, we have $\frac{\mathbf{54}}{6} = 9$.

20. The missing denominator is the number that we can divide 12 by to get 2. Since 12 ÷ 6 = 2, the denominator is 6:

$$\frac{12}{\mathbf{6}} = 2.$$

21. The missing denominator is the number that we can divide 56 by to get 7. Since 56 ÷ 8 = 7, the denominator is 8:

$$\frac{56}{\mathbf{8}} = 7.$$

22. The missing denominator is the number that we can divide 35 by to get 5. Since 35 ÷ 7 = 5, the denominator is 7:

$$\frac{35}{\mathbf{7}} = 5.$$

23. The missing denominator is the number that we can divide 36 by to get 4. Since 36 ÷ 9 = 4, the denominator is 9:

$$\frac{36}{\mathbf{9}} = 4.$$

We first evaluate each of the fractions, then connect.

24.
$\frac{6}{6} = 1$

$\frac{10}{5} = 2$

$\frac{9}{3} = 3$

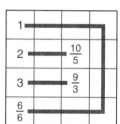

25.
$\frac{5}{5} = 1$

$\frac{4}{2} = 2$

$\frac{6}{2} = 3$

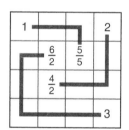

26.
$\frac{6}{3} = 2$

$\frac{12}{4} = 3$

$\frac{8}{2} = 4$

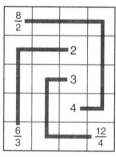

27.
$\frac{11}{11} = 1$

$\frac{12}{6} = 2$

$\frac{18}{6} = 3$

$\frac{40}{10} = 4$

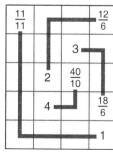

28.
$\frac{60}{30} = 2$

$\frac{15}{5} = 3$

$\frac{36}{9} = 4$

$\frac{42}{7} = 6$

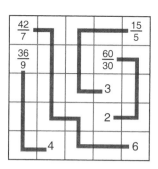

29.
$\frac{21}{7} = 3$

$\frac{32}{8} = 4$

$\frac{45}{9} = 5$

$\frac{80}{8} = 10$

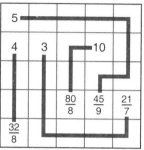

30.
$\frac{30}{15} = 2$

$\frac{20}{5} = 4$

$\frac{66}{11} = 6$

$\frac{56}{7} = 8$

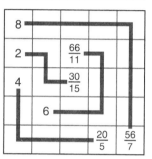

31.
$\frac{30}{10} = 3$

$\frac{75}{15} = 5$

$\frac{63}{9} = 7$

$\frac{72}{8} = 9$

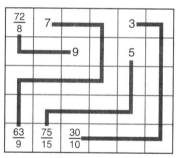

32.
$\frac{50}{25} = 2$

$\frac{16}{4} = 4$

$\frac{18}{3} = 6$

$\frac{54}{6} = 9$

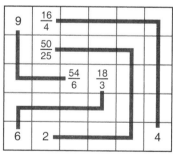

33.
$\frac{38}{19} = 2$

$\frac{24}{8} = 3$

$\frac{35}{7} = 5$

$\frac{84}{12} = 7$

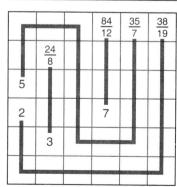

34. The tick marks split the number line between 0 and 1 into four equal pieces. Each piece has a length of $\frac{1}{4}$.
The arrow points to the end of the third piece, so we count 3 lengths of $\frac{1}{4}$ from 0 to $\frac{3}{4}$.

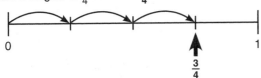

35. The tick marks split the number line between 0 and 1 into seven equal pieces. Each piece has a length of $\frac{1}{7}$.
The arrow points to the end of the second piece, so we count 2 lengths of $\frac{1}{7}$ from 0 to $\frac{2}{7}$.

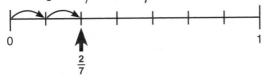

36. The tick marks split the number line between 0 and 1 into eight equal pieces. Each piece has a length of $\frac{1}{8}$.
The arrow points to the end of the third piece, so we count 3 lengths of $\frac{1}{8}$ from 0 to $\frac{3}{8}$.

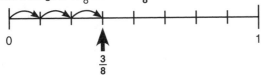

37. The tick marks split the number line between 0 and 1 into ten equal pieces. Each piece has a length of $\frac{1}{10}$.
The arrow points to the end of the seventh piece, so we count 7 lengths of $\frac{1}{10}$ from 0 to $\frac{7}{10}$.

38. The tick marks split the number line between 0 and 1 into five equal pieces. Each piece has a length of $\frac{1}{5}$.
The arrow points to the end of the third piece, so we count 3 lengths of $\frac{1}{5}$ from 0 to $\frac{3}{5}$.

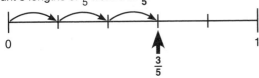

39. The tick marks split the number line between 0 and 1 into six equal pieces. Each piece has a length of $\frac{1}{6}$.
The arrow points to the end of the fifth piece, so we count 5 lengths of $\frac{1}{6}$ from 0 to $\frac{5}{6}$.

40. We use the marked number line to locate $\frac{3}{4}$ and $\frac{7}{9}$, as shown:

Since $\frac{3}{4}$ is to the left of $\frac{7}{9}$ on the number line, $\frac{3}{4}$ ⓐ $\frac{7}{9}$.

41. We use the marked number line to locate $\frac{3}{11}$ and $\frac{1}{3}$, as shown:

Since $\frac{3}{11}$ is to the left of $\frac{1}{3}$ on the number line, $\frac{3}{11}$ ⓐ $\frac{1}{3}$.

42. We use the marked number line to locate $\frac{6}{10}$ and $\frac{4}{7}$, as shown:

Since $\frac{6}{10}$ is to the right of $\frac{4}{7}$ on the number line, $\frac{6}{10}$ ⓑ $\frac{4}{7}$.

43. We use the marked number line to locate $\frac{2}{6}$ and $\frac{4}{12}$, as shown:

Since $\frac{2}{6}$ and $\frac{4}{12}$ are at the same point on the number line, $\frac{2}{6}$ ⓔ $\frac{4}{12}$.

44. We use the marked number line to locate $\frac{8}{13}$ and $\frac{2}{3}$, as shown:

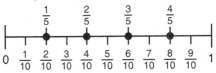

Since $\frac{8}{13}$ is to the left of $\frac{2}{3}$ on the number line, $\frac{8}{13}$ ⓐ $\frac{2}{3}$.

45. We mark points on the number line below that are represented by two equivalent fractions:

So, the four pairs of equivalent fractions on this line are $\frac{1}{5}=\frac{2}{10}$, $\frac{2}{5}=\frac{4}{10}$, $\frac{3}{5}=\frac{6}{10}$ and $\frac{4}{5}=\frac{8}{10}$.

46. The top part of the line has been split into three equal parts, and the bottom part of the line has been split into twelve equal parts. We label the tick marks with the numbers they represent on each side of the line. Then, we mark points on the number line that are represented by two equivalent fractions:

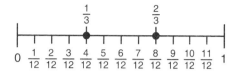

So, the two pairs of equivalent fractions on this line are $\frac{1}{3}=\frac{4}{12}$ and $\frac{2}{3}=\frac{8}{12}$.

47. The top part of the line has been split into sixteen equal parts, and the bottom part of the line has been split into four equal parts. We label the tick marks with the numbers they represent on each side of the line. Then, we mark points on the number line that are represented by two equivalent fractions:

So, the three pairs of equivalent fractions on this line are $\frac{4}{16}=\frac{1}{4}$, $\frac{8}{16}=\frac{2}{4}$, and $\frac{12}{16}=\frac{3}{4}$.

48. The number line between 0 and 1 is split into 6 equal pieces of length $\frac{1}{6}$. It takes 2 of these lengths to reach the point indicated by the dot. So the dot marks $\frac{2}{6}$.

If we make every *second* tick mark bold, as shown below, then the number line between 0 and 1 is split into three equal pieces of length $\frac{1}{3}$. So, the dot also marks $\frac{1}{3}$.

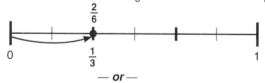

— *or* —

There are many other fractions that are equivalent to $\frac{2}{6}$. If we split the segment between each tick mark into two equal pieces, we see that the dot also marks $\frac{4}{12}$.

If we split the segment between each tick mark into three pieces, then we see that the dot also marks $\frac{6}{18}$. If we split the segment between each tick mark into four pieces, then we see that the dot also marks $\frac{8}{24}$, and so on.

49. The number line between 0 and 1 is split into 10 equal pieces of length $\frac{1}{10}$. It takes 8 of these lengths to reach the point indicated by the dot. So the dot marks $\frac{8}{10}$.

If we make every *second* tick mark bold, as shown below, then the number line between 0 and 1 is split into five equal pieces of length $\frac{1}{5}$. It takes 4 of these lengths to reach the point indicated by the dot. So, the dot also marks $\frac{4}{5}$.

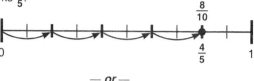

— *or* —

You might have split the segment between each tick mark into two, three, four, or any other number of equal pieces to find that $\frac{8}{10}$ is equal to $\frac{16}{20}, \frac{24}{30}, \frac{32}{40}$, and so on.

50. The number line between 0 and 1 is split into 10 equal pieces of length $\frac{1}{10}$. It takes 5 of these lengths to reach the point indicated by the dot. So the dot marks $\frac{5}{10}$.

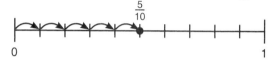

If we make every *fifth* tick mark bold, as shown below, then the number line between 0 and 1 is split into two equal pieces of length $\frac{1}{2}$. So, the dot also marks $\frac{1}{2}$.

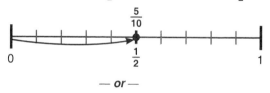

— *or* —

You might have split the segment between each tick mark into two, three, four, or any other number of equal pieces to find that $\frac{5}{10}$ is equal to $\frac{10}{20}, \frac{15}{30}, \frac{20}{40}$, and so on.

51. The number line between 0 and 1 is split into 9 equal pieces of length $\frac{1}{9}$. It takes 6 of these lengths to reach the point indicated by the dot. So the dot marks $\frac{6}{9}$.

If we make every *third* tick mark bold, as shown below, the number line between 0 and 1 is split into three equal pieces of length $\frac{1}{3}$. It takes 2 of these lengths to reach the point indicated by the dot. So, the dot also marks $\frac{2}{3}$.

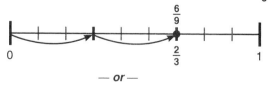

— *or* —

You might have split the segment between each tick mark into two, three, four, or any other number of equal pieces to find that $\frac{6}{9}$ is equal to $\frac{12}{18}, \frac{18}{27}, \frac{24}{36}$, and so on.

52. The tick marks split the number line between 3 and 4 into seven equal pieces, each of length $\frac{1}{7}$. One seventh more than $\frac{21}{7}$ is $\frac{22}{7}$, and $\frac{1}{7}$ more than $\frac{22}{7}$ is $\frac{23}{7}$, and so on. We label the rest of the sevenths between 3 and 4, as shown:

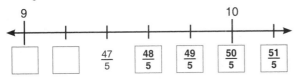

53. The tick marks split the number line between 9 and 10 into five equal pieces, each of length $\frac{1}{5}$. One fifth more than $\frac{47}{5}$ is $\frac{48}{5}$. We continue to count up by fifths to $\frac{50}{5} = 10$. The piece to the right of 10 also has length $\frac{1}{5}$, so the box on the far right marks the number that is one fifth more than $\frac{50}{5}$. One fifth more than $\frac{50}{5}$ is $\frac{51}{5}$.

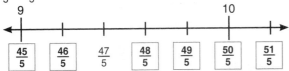

Then, one fifth less than $\frac{47}{5}$ is $\frac{46}{5}$, and one fifth less than $\frac{46}{5}$ is $\frac{45}{5} = 9$.

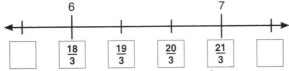

54. The tick marks split the number line between 6 and 7 into three equal pieces, each of length $\frac{1}{3}$. We write 6 and 7 as thirds: $6 = \frac{18}{3}$ and $7 = \frac{21}{3}$. Then, we label the remaining thirds between 6 and 7.

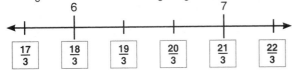

The piece to the left of 6 also has length $\frac{1}{3}$, so the box on the far left marks the number one third less than $\frac{18}{3}$. One third less than $\frac{18}{3}$ is $\frac{17}{3}$.

The piece to the right of 7 also has length $\frac{1}{3}$, so the box on the far right marks the number that is one third more than $\frac{21}{3}$. One third more than $\frac{21}{3}$ is $\frac{22}{3}$.

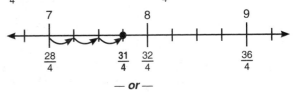

55. The tick marks split the number line between each pair of whole numbers into five equal pieces, each of length $\frac{1}{5}$.

We write 5, 6, and 7 as fifths: $5 = \frac{25}{5}$, $6 = \frac{30}{5}$, and $7 = \frac{35}{5}$. Then, we label these fractions on the given number line:

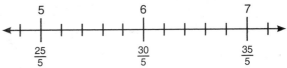

So, $\frac{31}{5}$ is between $\frac{30}{5} = 6$ and $\frac{35}{5} = 7$.
We count one fifth past $\frac{30}{5}$ on the number line to label $\frac{31}{5}$.

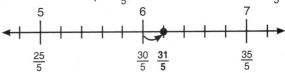

56. The tick marks split the number line between each pair of whole numbers into nine equal pieces, each of length $\frac{1}{9}$. We write 8, 9, and 10 as ninths: $8 = \frac{72}{9}$, $9 = \frac{81}{9}$, and $10 = \frac{90}{9}$. We label these fractions on the given number line:

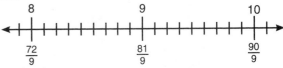

So, $\frac{85}{9}$ is between 9 and 10. We count four ninths past $\frac{81}{9}$ on the number line to mark $\frac{85}{9}$.

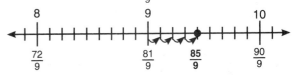

57. The tick marks split the number line between each pair of whole numbers into four equal pieces, each of length $\frac{1}{4}$.

We write 7, 8, and 9 as fourths: $7 = \frac{28}{4}$, $8 = \frac{32}{4}$, and $9 = \frac{36}{4}$. We label these fractions on the given number line:

So, $\frac{31}{4}$ is between 7 and 8. We count three fourths past $\frac{28}{4}$ on the number line to mark $\frac{31}{4}$.

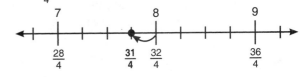

— *or* —

We count one fourth *less than* $\frac{32}{4}$ on the number line to mark $\frac{31}{4}$.

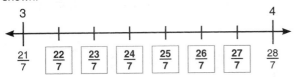

58. The tick marks split the number line between each pair of whole numbers into seven equal pieces, each of length $\frac{1}{7}$. The first whole number is three sevenths less than $\frac{24}{7}$, so it is $\frac{21}{7} = 3$.

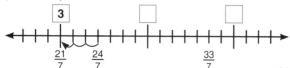

The second whole number is four sevenths more than $\frac{24}{7}$ (and five sevenths less than $\frac{33}{7}$), so it is $\frac{28}{7} = 4$.

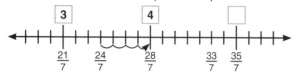

The third whole number is two sevenths more than $\frac{33}{7}$, so it is $\frac{35}{7} = 5$.

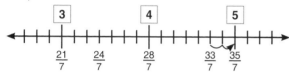

59. The tick marks split the number line between each pair of whole numbers into eight equal pieces, each of length $\frac{1}{8}$. The second whole number is three eighths less than $\frac{67}{8}$, so it is $\frac{64}{8} = 8$.

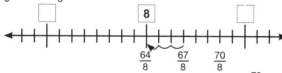

The third whole number is two eighths more than $\frac{70}{8}$, so it is $\frac{72}{8} = 9$.

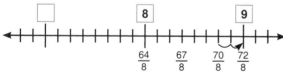

The first whole number is the number that is eight eighths ($\frac{8}{8} = 1$) less than $\frac{64}{8} = 8$. So, it is $\frac{56}{8} = 7$.

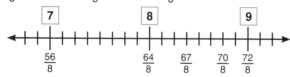

60. The tick marks split the number line between each pair of whole numbers into eight equal pieces, each of length $\frac{1}{8}$. We count three lengths of $\frac{1}{8}$ from 0 to reach the left arrow, so it marks $\frac{3}{8}$. We count 15 lengths of $\frac{1}{8}$ from 0 to reach the middle arrow, so it marks $\frac{15}{8}$. Finally, we count 21 lengths of $\frac{1}{8}$ from 0 to reach the right arrow, so it marks $\frac{21}{8}$.

— *or* —

We count three lengths of $\frac{1}{8}$ from 0 to reach the left arrow, so it marks $\frac{3}{8}$. The middle arrow is one eighth less than 2. Since $2 = \frac{16}{8}$, the middle arrow marks $\frac{15}{8}$. The third arrow is five eighths more than 2. Since $2 = \frac{16}{8}$, the third arrow marks $\frac{21}{8}$.

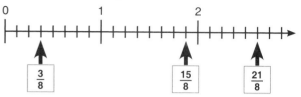

61. There are six possible ways to arrange the digits 1, 2, and 3 to create a fraction that is greater than 1. We organize these by denominator, then evaluate:

$\frac{23}{1}$ is equal to 23 ($23 \div 1 = 23$).

$\frac{32}{1}$ is equal to 32 ($32 \div 1 = 32$).

$\frac{13}{2}$ is between 6 and 7 ($12 \div 2 = 6$ and $14 \div 2 = 7$).

$\frac{31}{2}$ is between 15 and 16 ($30 \div 2 = 15$ and $32 \div 2 = 16$).

$\frac{12}{3}$ is equal to 4 ($12 \div 3 = 4$).

$\frac{21}{3}$ is equal to 7 ($21 \div 3 = 7$).

Of these fractions, only $\frac{31}{2}$ is between 15 and 16.

62. There are six possible ways to arrange the digits 4, 5, and 6 to create a fraction that is greater than 1. We organize these by denominator, then evaluate:

$\frac{56}{4}$ is equal to 14 ($56 \div 4 = 14$).

$\frac{65}{4}$ is between 16 and 17 ($64 \div 4 = 16$ and $68 \div 4 = 17$).

$\frac{46}{5}$ is between 9 and 10 ($45 \div 5 = 9$ and $50 \div 5 = 10$).

$\frac{64}{5}$ is between 12 and 13 ($60 \div 5 = 12$ and $65 \div 5 = 13$).

$\frac{45}{6}$ is between 7 and 8 ($42 \div 6 = 7$ and $48 \div 6 = 8$).

$\frac{54}{6}$ is equal to 9 ($54 \div 6 = 9$).

Of these fractions, only $\frac{45}{6}$ is between 7 and 8.

63. We know $\frac{20}{3}$ is between $\frac{18}{3} = 6$ and $\frac{21}{3} = 7$, so 7 is the smallest whole number that is greater than $\frac{20}{3}$. Then, we continue counting whole numbers:

$\frac{21}{3} = 7$, $\frac{24}{3} = 8$, $\frac{27}{3} = 9$, $\frac{30}{3} = 10$, $\frac{33}{3} = 11$, $\frac{36}{3} = 12$, $\frac{39}{3} = 13$.

The next-largest whole number, $\frac{42}{3} = 14$, is larger than $\frac{40}{3}$. So, the whole numbers between $\frac{20}{3}$ and $\frac{40}{3}$ are 7, 8, 9, 10, 11, 12, and 13. All together, that is **7** whole numbers.

— *or* —

We know that $\frac{20}{3}$ is between $\frac{18}{3} = 6$ and $\frac{21}{3} = 7$. We also know that $\frac{40}{3}$ is between $\frac{39}{3} = 13$ and $\frac{42}{3} = 14$. So, the whole numbers between $\frac{20}{3}$ and $\frac{40}{3}$ are 7, 8, 9, 10, 11, 12, and 13. All together, that is **7** whole numbers.

64. We know that $\frac{24}{6} = 4$.

$\frac{25}{6}$ is one sixth more than 4, so we write $\frac{25}{6}$ as $4\frac{1}{6}$.

$\frac{26}{6}$ is two sixths more than 4, so we write $\frac{26}{6}$ as $4\frac{2}{6}$.

$\frac{27}{6}$ is three sixths more than 4, so we write $\frac{27}{6}$ as $4\frac{3}{6}$.

$\frac{28}{6}$ is four sixths more than 4, so we write $\frac{28}{6}$ as $4\frac{4}{6}$.

$\frac{29}{6}$ is five sixths more than 4, so we write $\frac{29}{6}$ as $4\frac{5}{6}$.

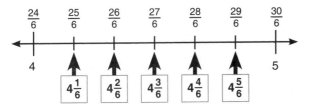

65. The left arrow marks $\frac{6}{5}$, which is one fifth more than $\frac{5}{5} = 1$. So, we label it $1\frac{1}{5}$. The middle arrow marks $\frac{9}{5}$, which is four fifths more than $\frac{5}{5} = 1$. So, we label it $1\frac{4}{5}$. The right arrow marks $\frac{12}{5}$, which is two fifths more than $\frac{10}{5} = 2$. So, we label it $2\frac{2}{5}$.

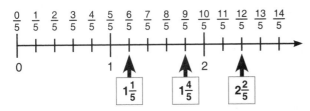

66. We start by locating and labeling the whole numbers on the given number line: $\frac{21}{3} = 7$, $\frac{24}{3} = 8$, and $\frac{27}{3} = 9$.

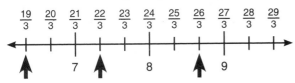

Since $\frac{22}{3}$ is one third more than 7, we label the middle arrow $7\frac{1}{3}$. Similarly, $\frac{26}{3}$ is two thirds more than 8. So, we label the right arrow $8\frac{2}{3}$. Although 6 is not shown on the number line, we know that $\frac{18}{3} = 6$. So, $\frac{19}{3}$ is one third more than 6. We label the left arrow $6\frac{1}{3}$.

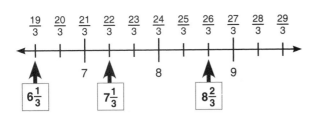

67. We first locate the whole numbers on the number line:

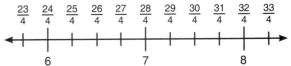

Then, we locate $6\frac{3}{4}$ by counting three fourths more than 6. Similarly, we locate $7\frac{1}{4}$ by counting one fourth more than 7.

Although 5 is not shown on the number line, we know that $5\frac{3}{4}$ is three fourths more than 5. Since $5 = \frac{20}{4}$, three fourths more than 5 is $\frac{23}{4}$.

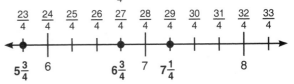

68. Since $\frac{19}{8}$ is three eighths more than $\frac{16}{8} = 2$, we have $\frac{19}{8} = 2\frac{3}{8}$.

69. Since $\frac{47}{6}$ is five sixths more than $\frac{42}{6} = 7$, we have $\frac{47}{6} = 7\frac{5}{6}$.

70. The tick marks split the number line between each pair of whole numbers into 7 equal pieces, so they mark sevenths on the line. The left arrow marks three sevenths more than 1, so we label it $1\frac{3}{7}$. The middle arrow marks five sevenths more than 2, so we label it $2\frac{5}{7}$. The right arrow marks two sevenths more than 3, so we label it $3\frac{2}{7}$.

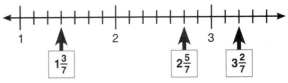

71. The tick marks split the number line into 5 equal pieces between each whole number, so they mark fifths on the line. The middle arrow marks two fifths more than 10, so we label it $10\frac{2}{5}$. The right arrow marks one fifth more than 11, so we label it $11\frac{1}{5}$.

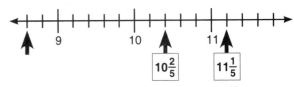

The left arrow marks two fifths less than $9 = \frac{45}{5}$, so the left arrow marks $\frac{43}{5}$. We know $\frac{40}{5} = 8$, so $\frac{43}{5}$ is three fifths more than 8. We label the left arrow $8\frac{3}{5}$.

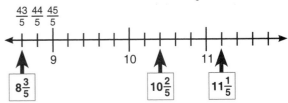

72. The tick marks split the number line into thirds. We label the missing fractions first and locate the whole numbers on the line.

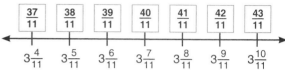

We know that $2 = \frac{6}{3}$. Since $\frac{8}{3}$ is two thirds more than 2, the left arrow marks $2\frac{2}{3}$. The middle arrow marks one third more than 3, so we label it $3\frac{1}{3}$. The right arrow marks two thirds more than 4, so we label it $4\frac{2}{3}$.

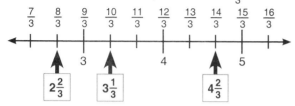

73. Since $\frac{11}{4}$ is three fourths more than $\frac{8}{4} = 2$, we have $\frac{11}{4} = 2\frac{3}{4}$. So, the recipe requires $2\frac{3}{4}$ cups of flour.
Since $\frac{3}{2}$ is one half more than $\frac{2}{2} = 1$, we have $\frac{3}{2} = 1\frac{1}{2}$. So, the recipe requires $1\frac{1}{2}$ cups of butter.
Since $\frac{7}{4}$ is three fourths more than $\frac{4}{4} = 1$, we have $\frac{7}{4} = 1\frac{3}{4}$. So, the recipe requires $1\frac{3}{4}$ cups of chocolate chips.
Since $\frac{5}{3}$ is two thirds more than $\frac{3}{3} = 1$, we have $\frac{5}{3} = 1\frac{2}{3}$. So, the recipe requires $1\frac{2}{3}$ cups of sugar.

The rewritten measurements are given below.

$2\frac{3}{4}$ cups flour	$1\frac{1}{2}$ cups butter
$1\frac{3}{4}$ cups chocolate chips	$1\frac{2}{3}$ cups sugar

74. The tick marks split the number line into sevenths. Since $5 = \frac{35}{7}$, we know $5\frac{1}{7} = \frac{36}{7}$. Each marked number is one seventh greater than the number to its left. We label the fractions as shown below:

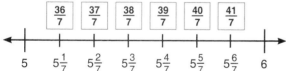

75. The tick marks split this number line into fifths. We label the whole number as a fraction with a denominator of five: $2 = \frac{10}{5}$.

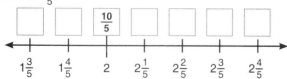

Each marked number is one fifth greater than the number to its left and one fifth less than the number to its right.

We label the remaining fractions as shown.

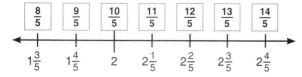

76. The tick marks split the number line into elevenths. We begin by labeling the smallest number on the line. We know that $3\frac{4}{11}$ is four elevenths more than 3.
Four elevenths more than $\frac{33}{11}$ is $\frac{37}{11}$. So, $3\frac{4}{11} = \frac{37}{11}$.
Each marked number is one eleventh greater than the number to its left, so we label the rest of the line.

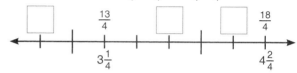

77. The tick marks split the number line into fourths. We begin by writing the labeled mixed numbers as fractions. We know that $3 = \frac{12}{4}$, and $3\frac{1}{4}$ is one fourth more than 3.
One fourth more than $\frac{12}{4}$ is $\frac{13}{4}$, so $3\frac{1}{4} = \frac{13}{4}$.
Similarly, $4 = \frac{16}{4}$, and $4\frac{2}{4}$ is two fourths more than 4.
Two fourths more than $\frac{16}{4}$ is $\frac{18}{4}$, so $4\frac{2}{4} = \frac{18}{4}$.

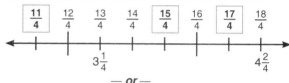

We use these numbers to label the boxed fractions:

— *or* —

We label the whole numbers on the number line:

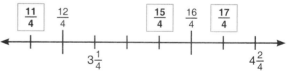

The left box marks one fourth less than $3 = \frac{12}{4}$. One fourth less than $\frac{12}{4}$ is $\frac{11}{4}$. The middle box marks three fourths more than 3. Three fourths more than $\frac{12}{4}$ is $\frac{15}{4}$.
The right box marks one fourth more than $4 = \frac{16}{4}$.
One fourth more than $\frac{16}{4}$ is $\frac{17}{4}$.

78. The rectangle is split into four equal pieces. So, each piece is $\frac{1}{4}$ of the whole rectangle. Three pieces are shaded, so $\frac{3}{4}$ of the rectangle is shaded.

79. The triangle is split into three equal pieces. So, each piece is $\frac{1}{3}$ of the whole triangle. One piece is shaded, so $\frac{1}{3}$ of the triangle is shaded.

80. The pentagon is split into five equal pieces. So, each piece is $\frac{1}{5}$ of the whole pentagon. Two pieces are shaded, so $\frac{2}{5}$ of the pentagon is shaded.

81. The rectangle is split into six equal pieces. So, each piece is $\frac{1}{6}$ of the whole rectangle. Five pieces are shaded, so $\frac{5}{6}$ of the rectangle is shaded.

82. The triangle is split into nine equal pieces. So, each piece is $\frac{1}{9}$ of the whole triangle. Four pieces are shaded, so $\frac{4}{9}$ of the triangle is shaded.

83. The shape is split into eleven equal pieces. So, each piece is $\frac{1}{11}$ of the whole shape. Six pieces are shaded, so $\frac{6}{11}$ of the shape is shaded.

84. The rectangle is split into eight equal pieces. So, each piece is $\frac{1}{8}$ of the rectangle. To shade $\frac{3}{8}$ of the rectangle, we shade any three of the eight pieces. For example,

 *or*

85. The decagon is split into ten equal pieces. So, each piece is $\frac{1}{10}$ of the decagon. To shade $\frac{7}{10}$ of the decagon, we shade any seven of the ten pieces. For example,

 or

86. The rectangle is split into five equal pieces. So, each piece is $\frac{1}{5}$ of the rectangle. To shade $\frac{2}{5}$ of the rectangle, we shade any two of the five pieces. For example,

 or

87. The triangle is split into six equal pieces. So, each piece is $\frac{1}{6}$ of the triangle. To shade $\frac{5}{6}$ of the triangle, we shade any five of the six pieces. For example,

 or

88. The shape is split into seven equal pieces. So, each piece is $\frac{1}{7}$ of the shape. To shade $\frac{4}{7}$ of the shape, we shade any four of the seven pieces. For example,

 or

89. The rectangle is split into twelve equal pieces. So, each piece is $\frac{1}{12}$ of the rectangle. To shade $\frac{5}{12}$ of the rectangle, we shade any five of the twelve pieces. For example,

90. The shape is split into nine equal pieces. So, each piece is $\frac{1}{9}$ of the shape. To shade $\frac{2}{9}$ of the shape, we shade any two of the nine pieces. For example,

 or

91. The shape is split into eight equal pieces. So, each piece is $\frac{1}{8}$ of the shape. To shade $\frac{5}{8}$ of the shape, we shade any five of the eight pieces. For example,

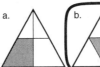 *or*

92. a. This triangle has been split into four pieces. One piece has been shaded, but the pieces are not the same size. This triangle is **not $\frac{1}{4}$ shaded**.

b. This triangle has been split into four equal pieces. One piece has been shaded, so this triangle **is $\frac{1}{4}$ shaded**.

c. This triangle has been split into four pieces. One piece has been shaded, but the pieces are not the same size. This triangle is **not $\frac{1}{4}$ shaded**.

d. This triangle has been split into six equal pieces. So, each piece is $\frac{1}{6}$ of the triangle. This triangle is $\frac{1}{6}$ shaded, so it is **not $\frac{1}{4}$ shaded**.

The only triangle below that is $\frac{1}{4}$ shaded is **b**.

93. The pentagon is split into five equal pieces. So, each piece is $\frac{1}{5}$ of the pentagon. To leave $\frac{2}{5}$ of the pentagon unshaded, we leave two of the five pieces unshaded. So, we shade $5-2=3$ of the 5 pieces. For example,

 or

When $\frac{3}{5}$ of the shape is shaded, $\frac{2}{5}$ is not shaded.

94. The shape is split into seven equal pieces. So, each piece is $\frac{1}{7}$ of the shape. To leave $\frac{3}{7}$ of the shape unshaded, we leave three of the seven pieces unshaded. So, we shade $7-3=4$ of the 7 pieces. For example,

When $\frac{4}{7}$ of the shape is shaded, $\frac{3}{7}$ is not shaded.

95. The triangle is split into six equal pieces. So, each piece is $\frac{1}{6}$ of the triangle. To leave $\frac{1}{6}$ of the triangle unshaded, we leave one of the six pieces unshaded. So, we shade $6 - 1 = 5$ of the 6 pieces. For example,

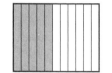 *or*

When $\frac{5}{6}$ of the shape is shaded, $\frac{1}{6}$ is not shaded.

96. The rectangle is split into eleven equal pieces. So, each piece is $\frac{1}{11}$ of the rectangle. To leave $\frac{6}{11}$ of the rectangle unshaded, we leave six of the eleven pieces unshaded. So, we shade $11 - 6 = 5$ of the 11 pieces. For example,

 or

When $\frac{5}{11}$ of the shape is shaded, $\frac{6}{11}$ is not shaded.

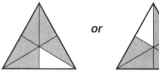

FRACTIONS
Equivalent Fractions   25-26

97. The rectangle is split into 16 equal pieces. From 16 pieces, we can make four groups, each with 4 pieces. Each group is $\frac{1}{4}$ of the whole rectangle. So, to shade $\frac{1}{4}$ of the rectangle, we shade one group. For example,

 *or*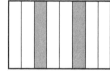

One group is four pieces or $\frac{4}{16}$ of the rectangle, so $\frac{4}{16} = \frac{1}{4}$.

98. The rectangle is split into 8 equal pieces. From 8 pieces, we can make four groups, each with 2 pieces. Each group is $\frac{1}{4}$ of the whole rectangle. So, to shade $\frac{1}{4}$ of the rectangle, we shade one group. For example,

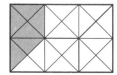 *or*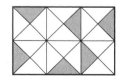

One group is two pieces or $\frac{2}{8}$ of the rectangle, so $\frac{2}{8} = \frac{1}{4}$.

99. The rectangle is split into 24 equal pieces. From 24 pieces, we can make four groups, each with 6 pieces. Each group is $\frac{1}{4}$ of the whole rectangle. So, to shade $\frac{1}{4}$ of the rectangle, we shade one group of six. For example,

One group is six pieces or $\frac{6}{24}$ of the rectangle, so $\frac{6}{24} = \frac{1}{4}$.

100. The triangle is split into 6 equal pieces. From 6 pieces, we can make two groups, each with 3 pieces. Each group is $\frac{1}{2}$ of the whole triangle. So, to shade $\frac{1}{2}$ of the triangle, we shade one group of three pieces. This is $\frac{3}{6}$ of the triangle. For example,

 or

101. The circle is split into 10 equal pieces. From 10 pieces, we can make five groups, each with 2 pieces. Each group is $\frac{1}{5}$ of the whole circle. So, to shade $\frac{1}{5}$ of the circle, we shade one group of two pieces. This is $\frac{2}{10}$ of the circle. For example,

 or

102. The shape is split into 6 equal pieces. From 6 pieces, we can make three groups, each with 2 pieces. Each group is $\frac{1}{3}$ of the whole shape. So, to shade $\frac{1}{3}$ of the shape, we shade one group of two pieces. This is $\frac{2}{6}$ of the shape. For example,

 or

103. The hexagon is split into 12 equal pieces. From 12 pieces, we can make six groups, each with 2 pieces. Each group is $\frac{1}{6}$ of the whole hexagon. So, to shade $\frac{5}{6}$ of the hexagon, we shade five groups of two. This is $5 \times 2 = 10$ pieces or $\frac{10}{12}$ of the hexagon. For example,

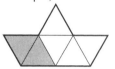 *or*

104. The shape is split into 10 equal pieces. From 10 pieces, we can make five groups, each with 2 pieces. Each group is $\frac{1}{5}$ of the whole shape. So, to shade $\frac{2}{5}$ of the shape, we shade two groups of two. This is $2 \times 2 = 4$ pieces or $\frac{4}{10}$ of the shape. For example,

 or

105. The shape is split into 8 equal pieces. From 8 pieces, we can make four groups, each with 2 pieces. Each group is $\frac{1}{4}$ of the whole shape. So, to shade $\frac{3}{4}$ of the shape, we shade three groups of two. This is $3 \times 2 = 6$ pieces or $\frac{6}{8}$ of the shape. For example,

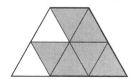 *or*

106. a. This shape has been split into four equal pieces. One piece has been shaded, so this shape **is $\frac{1}{4}$ shaded**.

b. This shape has been split into twelve equal pieces. From 12 pieces, we can make four groups, each with 3 pieces. Each group is $\frac{1}{4}$ of the whole shape. Three pieces have been shaded, so this shape **is $\frac{1}{4}$ shaded**.

c. This shape has been split into four pieces. One piece has been shaded, but the pieces are not the same size. This shape **is not $\frac{1}{4}$ shaded**.

d. This shape has been split into twelve equal pieces. From 12 pieces, we can make four groups, each with 3 pieces. Each group is $\frac{1}{4}$ of the whole shape. Three pieces have been shaded, so this shape **is $\frac{1}{4}$ shaded**.

The only shape that is not $\frac{1}{4}$ shaded is **c**.

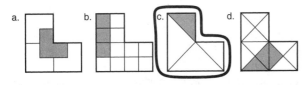

107. $4 \times 2 = 8$, and $9 \times 2 = 18$.
$$\frac{4}{9} \overset{\times 2}{\underset{\times 2}{=}} \frac{\mathbf{8}}{\mathbf{18}}$$

108. $2 \times 2 = 4$, and $7 \times 2 = 14$.
$$\frac{2}{7} \overset{\times 2}{\underset{\times 2}{=}} \frac{\mathbf{4}}{\mathbf{14}}$$

109. $4 \times 4 = 16$, and $3 \times 4 = 12$.
$$\frac{4}{3} \overset{\times 4}{\underset{\times 4}{=}} \frac{\mathbf{16}}{\mathbf{12}}$$

110. $7 \times 5 = 35$, and $10 \times 5 = 50$.
$$\frac{7}{10} \overset{\times 5}{\underset{\times 5}{=}} \frac{\mathbf{35}}{\mathbf{50}}$$

111. $6 \div 3 = 2$, and $9 \div 3 = 3$.
$$\frac{6}{9} \overset{\div 3}{\underset{\div 3}{=}} \frac{\mathbf{2}}{\mathbf{3}}$$

112. $15 \div 5 = 3$, and $25 \div 5 = 5$.
$$\frac{15}{25} \overset{\div 5}{\underset{\div 5}{=}} \frac{\mathbf{3}}{\mathbf{5}}$$

113. $36 \div 4 = 9$, and $28 \div 4 = 7$.
$$\frac{36}{28} \overset{\div 4}{\underset{\div 4}{=}} \frac{\mathbf{9}}{\mathbf{7}}$$

114. $18 \div 9 = 2$, and $81 \div 9 = 9$.
$$\frac{18}{81} \overset{\div 9}{\underset{\div 9}{=}} \frac{\mathbf{2}}{\mathbf{9}}$$

115. $3 \times \boxed{10} = 30$. So, to make an equivalent fraction with denominator 30, we multiply both the numerator and denominator of $\frac{2}{3}$ by 10:
$$\frac{2}{3} \overset{\times 10}{\underset{\times 10}{=}} \frac{\mathbf{20}}{\mathbf{30}}$$

116. $4 \div \boxed{4} = 1$. So, to make an equivalent fraction with numerator 1, we divide both the numerator and denominator of $\frac{4}{12}$ by 4:
$$\frac{4}{12} \overset{\div 4}{\underset{\div 4}{=}} \frac{\mathbf{1}}{\mathbf{3}}$$

117. $14 \div \boxed{2} = 7$. So, to make an equivalent fraction with denominator 7, we divide both the numerator and denominator of $\frac{2}{14}$ by 2:
$$\frac{2}{14} \overset{\div 2}{\underset{\div 2}{=}} \frac{\mathbf{1}}{\mathbf{7}}$$

118. $6 \times \boxed{2} = 12$. So, to make an equivalent fraction with denominator 12, we multiply both the numerator and denominator of $\frac{5}{6}$ by 2:
$$\frac{5}{6} \overset{\times 2}{\underset{\times 2}{=}} \frac{\mathbf{10}}{\mathbf{12}}$$

119. $6 \times \boxed{5} = 30$. So, to make an equivalent fraction with numerator 30, we multiply both the numerator and denominator of $\frac{6}{7}$ by 5:
$$\frac{6}{7} \overset{\times 5}{\underset{\times 5}{=}} \frac{\mathbf{30}}{\mathbf{35}}$$

120. $15 \div \boxed{3} = 5$. So, to make an equivalent fraction with denominator 5, we divide both the numerator and denominator of $\frac{12}{15}$ by 3:
$$\frac{12}{15} \overset{\div 3}{\underset{\div 3}{=}} \frac{\mathbf{4}}{\mathbf{5}}$$

121. $8 \times \boxed{4} = 32$. So, to make an equivalent fraction with denominator 32, we multiply both the numerator and denominator of $\frac{2}{8}$ by 4:
$$\frac{2}{8} \overset{\times 4}{\underset{\times 4}{=}} \frac{\mathbf{8}}{\mathbf{32}}$$

122. $7 \times \boxed{6} = 42$. So, to make an equivalent fraction with denominator 42, we multiply both the numerator and denominator of $\frac{6}{7}$ by 6:
$$\frac{6}{7} \overset{\times 6}{\underset{\times 6}{=}} \frac{\mathbf{36}}{\mathbf{42}}$$

123. $12 \div \boxed{6} = 2$. So, to make an equivalent fraction with numerator 2, we divide both the numerator and denominator of $\frac{12}{78}$ by 6:
$$\frac{12}{78} \overset{\div 6}{\underset{\div 6}{=}} \frac{\mathbf{2}}{\mathbf{13}}$$

124. $20 \times \boxed{4} = 80$. So, to make an equivalent fraction with denominator 80, we multiply both the numerator and denominator of $\frac{13}{20}$ by 4:
$$\frac{13}{20} \overset{\times 4}{\underset{\times 4}{=}} \frac{\mathbf{52}}{\mathbf{80}}$$

125. We consider one pair of fractions at a time. Since all five fractions are equivalent, any two of these fractions are equal. We start with the first pair: $\frac{2}{5} = \frac{2}{10}$.

$5 \times \boxed{2} = 10$. So, to make an equivalent fraction with denominator 10, we multiply both the numerator and denominator of $\frac{2}{5}$ by 2:

$$\frac{2}{5} \xrightarrow[\times 2]{\times 2} \frac{\mathbf{4}}{10}$$

Next, we consider the third equivalent fraction. Since 6 is a multiple of 2, we look at the pair $\frac{2}{5} = \frac{6}{}$.

$2 \times \boxed{3} = 6$. So, to make an equivalent fraction with numerator 6, we multiply both the numerator and denominator of $\frac{2}{5}$ by 3:

$$\frac{2}{5} \xrightarrow[\times 3]{\times 3} \frac{6}{\mathbf{15}}$$

Next, we consider the fourth equivalent fraction. Since 30 is a multiple of 2, we look at the pair $\frac{2}{5} = \frac{30}{}$.

$2 \times \boxed{15} = 30$. So, to make an equivalent fraction with numerator 30, we multiply both the numerator and denominator of $\frac{2}{5}$ by 15:

$$\frac{2}{5} \xrightarrow[\times 15]{\times 15} \frac{30}{\mathbf{75}}$$

Finally, we consider the fifth equivalent fraction. Since 150 is a multiple of 2, we look at the pair $\frac{2}{5} = \frac{150}{}$.

$2 \times \boxed{75} = 150$. So, to make an equivalent fraction with numerator 150, we multiply both the numerator and denominator of $\frac{2}{5}$ by 75:

$$\frac{2}{5} \xrightarrow[\times 75]{\times 75} \frac{150}{\mathbf{375}}$$

All together, we have

$$\frac{2}{5} = \frac{\mathbf{4}}{10} = \frac{6}{\mathbf{15}} = \frac{30}{\mathbf{75}} = \frac{150}{\mathbf{375}}.$$

You may have compared other pairs of fractions to get the same answers.

126. Since all three fractions are equivalent, any two of these fractions are equal. We begin with the first pair of fractions: $\frac{1}{3} = \frac{4}{m}$.

$1 \times \boxed{4} = 4$, so to make an equivalent fraction with numerator 4, we multiply both the numerator and denominator of $\frac{1}{3}$ by 4:

$$\frac{1}{3} \xrightarrow[\times 4]{\times 4} \frac{4}{\mathbf{12}}$$

So, $m = \mathbf{12}$.
Next, we substitute $m = 12$ into the original equation:

$$\frac{1}{3} = \frac{4}{12} = \frac{\mathbf{12}}{n}$$

Then, we consider the second two fractions: $\frac{4}{12} = \frac{12}{n}$.
$4 \times \boxed{3} = 12$, so to make an equivalent fraction with numerator 12, we multiply both the numerator and denominator of $\frac{4}{12}$ by 3:

$$\frac{4}{12} \xrightarrow[\times 3]{\times 3} \frac{12}{\mathbf{36}}$$

So, $n = \mathbf{36}$.
All together, we have $\frac{1}{3} = \frac{4}{12} = \frac{12}{36}$.

127. $1 \times \boxed{s} = s$. So, to make an equivalent fraction with numerator s, we multiply both the numerator and denominator of $\frac{1}{s}$ by s:

$$\frac{1}{s} \xrightarrow[\times s]{\times s} \frac{s}{16}$$

So, $s \times s = 16$. Since $4 \times 4 = 16$, we have $s = \mathbf{4}$.

We substitute $s = 4$ into the original equation to check our answer:

$$\frac{1}{4} \xrightarrow[\times 4]{\times 4} \frac{4}{16} \quad \checkmark$$

128. We do not need to know the values of a and b to solve this problem. $3 \times \boxed{3} = 9$, so to make an equivalent fraction with numerator 9, we multiply both the numerator and denominator of $\frac{3}{a}$ by 3:

$$\frac{3}{a} \xrightarrow[\times 3]{\times 3} \frac{9}{b}$$

So, we see that $a \times 3 = b$.
Next, we look at the second pair of equivalent fractions:

$$\frac{5}{a} = \frac{x}{b}$$

Since $a \times 3 = b$, we multiply the numerator and denominator of any fraction with denominator a by 3 to get an equivalent fraction with denominator b:

$$\frac{5}{a} \xrightarrow[\times 3]{\times 3} \frac{\mathbf{15}}{b}$$

So, $x = \mathbf{15}$.

You may find it helpful to look at some values of a (such as $a = 1$ or $a = 7$) to see that x is *always* 15.

129. We simplify $\frac{3}{6}$ to an equivalent unit fraction:

$$\frac{3}{6} \xrightarrow[\div 3]{\div 3} \frac{1}{2}$$

Since $\frac{1}{2}$ is greater than $\frac{1}{4}$, we know $\frac{3}{6}$ is greater than $\frac{1}{4}$. So, $\frac{3}{6}$ **is not equivalent to** $\frac{1}{4}$.

130. No. As we saw in the previous problem, **adding the same number to the numerator and denominator does not always create an equivalent fraction.**

Adding to the numerator and denominator is *not* a valid way to convert a fraction.

The *only* time adding the same number to the numerator and denominator creates an equivalent fraction is when the numerator and denominator are the same.

131. We write $\frac{6}{5}$ as a mixed number: $1\frac{1}{5}$.
We also write $\frac{4}{3}$ as a mixed number: $1\frac{1}{3}$.
Since $\frac{1}{5}$ is less than $\frac{1}{3}$, we know $1\frac{1}{5}$ is less than $1\frac{1}{3}$. So, $\frac{6}{5}$ is less than $\frac{4}{3}$.

Therefore, $\frac{4}{3}$ **is not equivalent to** $\frac{6}{5}$.

132. No. As we saw in the previous problem, **subtracting the same number from the numerator and denominator does not always create an equivalent fraction**.

Subtracting from the numerator and denominator is *not* a valid way to convert a fraction. The *only* time subtracting the same number from the numerator and denominator creates an equivalent fraction is when the numerator and denominator are the same.

There are often several ways to simplify the fractions below to arrive at the answers given.

133. Both 6 and 8 are multiples of 2, so

$$\frac{6}{8} \overset{\div 2}{\underset{\div 2}{=}} \frac{3}{4}$$

Since 1 is the only whole number that divides each of 3 and 4 with no remainder, $\frac{3}{4}$ is in simplest form.

134. Both 9 and 18 are multiples of 3, so

$$\frac{9}{18} \overset{\div 3}{\underset{\div 3}{=}} \frac{3}{6}$$

Then, $\frac{3}{6}$ can be simplified. Both 3 and 6 are multiples of 3, so

$$\frac{3}{6} \overset{\div 3}{\underset{\div 3}{=}} \frac{1}{2}$$

Since 1 is the only whole number that divides each of 1 and 2 with no remainder, $\frac{1}{2}$ is in simplest form.

— *or* —

We simplify $\frac{9}{18}$ in one step by dividing each of 9 and 18 by 9:

$$\frac{9}{18} \overset{\div 9}{\underset{\div 9}{=}} \frac{1}{2}$$

Since 1 is the only whole number that divides each of 1 and 2 with no remainder, $\frac{1}{2}$ is in simplest form.

135. Both 15 and 45 are multiples of 5, so

$$\frac{15}{45} \overset{\div 5}{\underset{\div 5}{=}} \frac{3}{9}$$

Then, $\frac{3}{9}$ can be simplified. Both 3 and 9 are multiples of 3, so

$$\frac{3}{9} \overset{\div 3}{\underset{\div 3}{=}} \frac{1}{3}$$

Since 1 is the only whole number that divides each of 1 and 3 with no remainder, $\frac{1}{3}$ is in simplest form.

— *or* —

We simplify $\frac{15}{45}$ in one step by dividing each of 15 and 45 by 15:

$$\frac{15}{45} \overset{\div 15}{\underset{\div 15}{=}} \frac{1}{3}$$

Since 1 is the only whole number that divides each of 1 and 3 with no remainder, $\frac{1}{3}$ is in simplest form.

136. Both 22 and 40 are multiples of 2, so

$$\frac{22}{40} \overset{\div 2}{\underset{\div 2}{=}} \frac{11}{20}$$

Since 1 is the only whole number that divides each of 11 and 20 with no remainder, $\frac{11}{20}$ is in simplest form.

137. Both 12 and 18 are multiples of 2, so

$$\frac{12}{18} \overset{\div 2}{\underset{\div 2}{=}} \frac{6}{9}$$

Then, $\frac{6}{9}$ can be simplified. Both 6 and 9 are multiples of 3, so

$$\frac{6}{9} \overset{\div 3}{\underset{\div 3}{=}} \frac{2}{3}$$

Since 1 is the only whole number that divides each of 2 and 3 with no remainder, $\frac{2}{3}$ is in simplest form.

— *or* —

We simplify $\frac{12}{18}$ in one step by dividing each of 12 and 18 by 6:

$$\frac{12}{18} \overset{\div 6}{\underset{\div 6}{=}} \frac{2}{3}$$

Since 1 is the only whole number that divides each of 2 and 3 with no remainder, $\frac{2}{3}$ is in simplest form.

138. Since 1 is the only whole number that divides each of 13 and 24 with no remainder, $\frac{13}{24}$ is already in simplest form.

139. Both 32 and 56 are multiples of 2, so

$$\frac{32}{56} \overset{\div 2}{\underset{\div 2}{=}} \frac{16}{28}$$

Then, $\frac{16}{28}$ can be simplified. Both 16 and 28 are multiples of 2, so

$$\frac{16}{28} \overset{\div 2}{\underset{\div 2}{=}} \frac{8}{14}$$

Then, $\frac{8}{14}$ can be simplified. Both 8 and 14 are multiples of 2, so

$$\frac{8}{14} \overset{\div 2}{\underset{\div 2}{=}} \frac{4}{7}$$

Since 1 is the only whole number that divides each of 4 and 7 with no remainder, $\frac{4}{7}$ is in simplest form.

— *or* —

We simplify $\frac{32}{56}$ in one step by dividing each of 32 and 56 by 8:

$$\frac{32}{56} \overset{\div 8}{\underset{\div 8}{=}} \frac{4}{7}$$

Since 1 is the only whole number that divides each of 4 and 7 with no remainder, $\frac{4}{7}$ is in simplest form.

140. Both 72 and 60 are multiples of 2, so

$$\frac{72}{60} \overset{\div 2}{\underset{\div 2}{=}} \frac{36}{30}$$

Then, $\frac{36}{30}$ can be simplified. Both 36 and 30 are multiples of 3, so

$$\frac{36}{30} \overset{\div 3}{\underset{\div 3}{=}} \frac{12}{10}$$

Then, $\frac{12}{10}$ can be simplified. Both 12 and 10 are multiples of 2, so

$$\frac{12}{10} \overset{\div 2}{\underset{\div 2}{=}} \frac{6}{5}$$

Since 1 is the only whole number that divides each of 6 and 5 with no remainder, $\frac{\mathbf{6}}{\mathbf{5}}$ is in simplest form.

— *or* —

We simplify $\frac{72}{60}$ in one step by dividing each of 60 and 72 by 12:

$$\frac{72}{60} \overset{\div 12}{\underset{\div 12}{=}} \frac{6}{5}$$

Since 1 is the only whole number that divides each of 6 and 5 with no remainder, $\frac{\mathbf{6}}{\mathbf{5}}$ is in simplest form.

You may have also written $\frac{6}{5}$ as a mixed number in simplest form: $\mathbf{1\frac{1}{5}}$.

141. The tick marks split the number line between 0 and 1 into eight equal pieces, each with length $\frac{1}{8}$. The marked point is at $\frac{4}{8}$. However, $\frac{4}{8}$ can be simplified:

$$\frac{4}{8} \overset{\div 4}{\underset{\div 4}{=}} \frac{1}{2}$$

We could also bold every fourth tick mark as shown below. Then, the number line between 0 and 1 is split into two equal pieces of length $\frac{1}{2}$.

So, in simplest form, the marked point is at $\frac{\mathbf{1}}{\mathbf{2}}$.

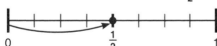

142. The tick marks split the number line between 0 and 1 into twelve equal pieces, each with length $\frac{1}{12}$. The marked point is at $\frac{4}{12}$. However, $\frac{4}{12}$ can be simplified:

$$\frac{4}{12} \overset{\div 4}{\underset{\div 4}{=}} \frac{1}{3}$$

We could also bold every fourth tick mark as shown below. Then, the number line between 0 and 1 is split into three equal pieces of length $\frac{1}{3}$.

So, in simplest form, the marked point is at $\frac{\mathbf{1}}{\mathbf{3}}$.

143. The tick marks split the number line between 0 and 1 into nine equal pieces, each with length $\frac{1}{9}$. The marked point is at $\frac{6}{9}$. However, $\frac{6}{9}$ can be simplified:

$$\frac{6}{9} \overset{\div 3}{\underset{\div 3}{=}} \frac{2}{3}$$

We could also bold every third tick mark, as shown below. Then, the number line between 0 and 1 is split into three equal pieces of length $\frac{1}{3}$.

So, in simplest form, the marked point is at $\frac{\mathbf{2}}{\mathbf{3}}$.

144. The tick marks split the number line between 0 and 1 into eight equal pieces, each with length $\frac{1}{8}$. The marked point is at $\frac{6}{8}$. However, $\frac{6}{8}$ can be simplified:

$$\frac{6}{8} \overset{\div 2}{\underset{\div 2}{=}} \frac{3}{4}$$

We could also bold every second tick mark, as shown below. Then, the number line between 0 and 1 is split into four equal pieces of length $\frac{1}{4}$.

So, in simplest form, the marked point is at $\frac{\mathbf{3}}{\mathbf{4}}$.

145. The tick marks split the number line between 0 and 1 into fifteen equal pieces, each with length $\frac{1}{15}$. The marked point is at $\frac{12}{15}$. However, $\frac{12}{15}$ can be simplified:

$$\frac{12}{15} \overset{\div 3}{\underset{\div 3}{=}} \frac{4}{5}$$

We could also bold every third tick mark, as shown below. Then, the number line between 0 and 1 is split into five equal pieces of length $\frac{1}{5}$.

So, in simplest form, the marked point is at $\frac{\mathbf{4}}{\mathbf{5}}$.

146. The tick marks split the number line between 0 and 1 and between 1 and 2 into eight equal pieces, each with length $\frac{1}{8}$. The marked point is at $\frac{10}{8}$. However, $\frac{10}{8}$ can be simplified:

$$\frac{10}{8} \overset{\div 2}{\underset{\div 2}{=}} \frac{5}{4}$$

We could also bold every second tick mark, as shown below. Then, the number line between 0 and 1 and between 1 and 2 is split into four equal pieces of length $\frac{1}{4}$. So, in simplest form, the marked point is at $\frac{\mathbf{5}}{\mathbf{4}}$.

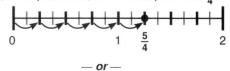

— *or* —

You may have written $\frac{5}{4}$ as a mixed number in simplest form and labeled the marked point $1\frac{1}{4}$.

147. The circle is split into twelve equal pieces, and three pieces are shaded. So, $\frac{3}{12}$ of the circle is shaded. However, $\frac{3}{12}$ can be simplified:

$$\frac{3}{12} \overset{\div 3}{\underset{\div 3}{=}} \frac{1}{4}$$

So, in simplest form, $\frac{1}{4}$ of the circle is shaded.

148. The shape is split into fourteen equal pieces, and four pieces are shaded. So, $\frac{4}{14}$ of the shape is shaded. However, $\frac{4}{14}$ can be simplified:

$$\frac{4}{14} \overset{\div 2}{\underset{\div 2}{=}} \frac{2}{7}$$

So, in simplest form, $\frac{2}{7}$ of the shape is shaded.

149. The triangle is split into sixteen equal pieces, and eight pieces are shaded. So, $\frac{8}{16}$ of the triangle is shaded. However, $\frac{8}{16}$ can be simplified:

$$\frac{8}{16} \overset{\div 8}{\underset{\div 8}{=}} \frac{1}{2}$$

So, in simplest form, $\frac{1}{2}$ of the triangle is shaded.

150. The rectangle is split into twelve equal pieces, and eight pieces are shaded. So, $\frac{8}{12}$ of the rectangle is shaded. However, $\frac{8}{12}$ can be simplified:

$$\frac{8}{12} \overset{\div 4}{\underset{\div 4}{=}} \frac{2}{3}$$

So, in simplest form, $\frac{2}{3}$ of the rectangle is shaded.

FRACTIONS
Constellation Puzzles   34-35

We first find sets of equivalent fractions in each puzzle. Writing each fraction in simplest form may help you to discover sets of equivalent fractions. The given solutions are the only correct ways to connect the dots.

151. In this puzzle, we have $\frac{1}{3}=\frac{4}{12}=\frac{8}{24}$ and $\frac{1}{2}=\frac{2}{4}=\frac{3}{6}$.

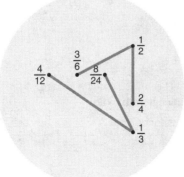

152. In this puzzle, we have $\frac{1}{7}=\frac{5}{35}=\frac{9}{63}$ and $\frac{1}{5}=\frac{5}{25}=\frac{8}{40}$.

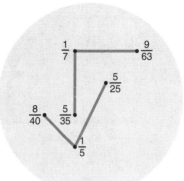

153. In this puzzle, we have the following equivalent fractions:

$$\frac{4}{5}=\frac{16}{20}$$
$$\frac{3}{8}=\frac{6}{16}$$
$$\frac{5}{12}=\frac{15}{36}$$

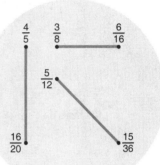

154. In this puzzle, we have the following equivalent fractions:

$$\frac{2}{5}=\frac{6}{15}=\frac{8}{20}$$
$$\frac{7}{10}=\frac{21}{30}=\frac{35}{50}$$
$$\frac{4}{7}=\frac{8}{14}=\frac{12}{21}$$

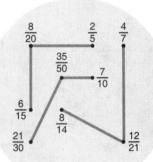

155. In this puzzle, we have the following equivalent fractions:

$$\frac{1}{4}=\frac{5}{20}=\frac{12}{48}$$
$$\frac{1}{12}=\frac{3}{36}=\frac{5}{60}$$
$$\frac{1}{9}=\frac{3}{27}=\frac{6}{54}$$

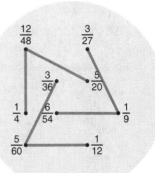

156. In this puzzle, we have the following equivalent fractions:

$\frac{8}{3}=\frac{32}{12}=\frac{48}{18}$

$\frac{7}{6}=\frac{35}{30}=\frac{63}{54}$

$\frac{4}{9}=\frac{12}{27}=\frac{20}{45}$

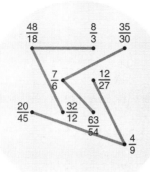

157. In this puzzle, we have the following equivalent fractions:

$\frac{2}{7}=\frac{4}{14}=\frac{14}{49}$

$\frac{4}{3}=\frac{20}{15}=\frac{32}{24}$

$\frac{5}{4}=\frac{25}{20}=\frac{45}{36}$

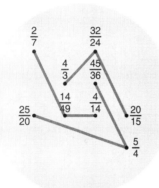

158. In this puzzle, we have the following equivalent fractions:

$\frac{13}{6}=\frac{26}{12}=\frac{39}{18}$

$\frac{11}{9}=\frac{22}{18}=\frac{33}{27}$

$\frac{10}{7}=\frac{40}{28}=\frac{70}{49}$

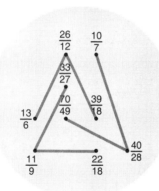

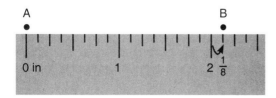

159. With careful measuring, we see that the distance from point A to point B is two whole inches, plus $\frac{1}{8}$ of an inch.

So the distance from point A to point B is $2\frac{1}{8}$ **inches**.

160. With careful measuring, we see that the distance from point B to point C is two whole inches, plus $\frac{4}{8}$ of an inch. We simplify $\frac{4}{8}$ to $\frac{1}{2}$. So, the distance from point B to point C is $2\frac{1}{2}$ **inches**.

161. We measure the distance from point A to point C with a ruler to find the distance is $4\frac{5}{8}$ **inches**.

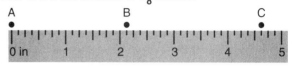

— or —

The distance from A to C is the sum of the distance from A to B and the distance from B to C. The distance from A to B is $2\frac{1}{8}$ inches. The distance from B to C is $2\frac{4}{8}$ inches. So, the distance from A to C is $2\frac{1}{8}+2\frac{4}{8}$ inches. Since $2\frac{1}{8}=2+\frac{1}{8}$ and $2\frac{4}{8}=2+\frac{4}{8}$, we have

$$2\frac{1}{8}+2\frac{4}{8}=2+\frac{1}{8}+2+\frac{4}{8}$$

$$=2+2+\frac{1}{8}+\frac{4}{8}$$

We first add the whole numbers and then the fractions: $2+2$ is 4, and $\frac{1}{8}+\frac{4}{8}$ is $\frac{5}{8}$.

So, $(2+2)+\left(\frac{1}{8}+\frac{4}{8}\right)=4+\frac{5}{8}=4\frac{5}{8}$.

So the distance from point A to point C is $4\frac{5}{8}$ **inches**.

162. The distance from point D to point B is $1\frac{1}{8}$. The greatest possible distance from A to D occurs when points A, B, and D are all in a line, with point D to the right of point B. (To see why, you might review the Triangle Inequality section in Chapter 3 of Practice 3A.) We measure $1\frac{1}{8}$ inches to the right of B and place D as shown:

Then, we measure the distance from point A to point D: $3\frac{2}{8}$ inches. We simplify $\frac{2}{8}$ to $\frac{1}{4}$. In simplest form, the distance from A to D is $3\frac{1}{4}$ inches.

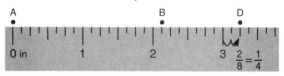

— or —

Once we have placed D, the distance from point A to point D is the sum of the distance from A to B and the distance from B to D.

The distance from A to B is $2\frac{1}{8}$ inches. The distance from B to D is $1\frac{1}{8}$ inches. So, the distance from point A to point D is $2\frac{1}{8}+1\frac{1}{8}$ inches.

Since $2\frac{1}{8}=2+\frac{1}{8}$ and $1\frac{1}{8}=1+\frac{1}{8}$, we have
$$2\frac{1}{8}+1\frac{1}{8}=2+\frac{1}{8}+1+\frac{1}{8}$$
$$=2+1+\frac{1}{8}+\frac{1}{8}$$

We first add the whole numbers and then the fractions: $2+1$ is 3, and $\frac{1}{8}+\frac{1}{8}$ is $\frac{2}{8}$.

So, $(2+1)+(\frac{1}{8}+\frac{1}{8})=3+\frac{2}{8}=3\frac{2}{8}$.

We simplify $\frac{2}{8}$ to $\frac{1}{4}$. So, in simplest form, the greatest possible distance from A to D is $3\frac{1}{4}$ **inches**.

163. The distance from point D to point B is $1\frac{1}{8}$ inches. The smallest distance between points A and D occurs when A, B, and D are all in a line, with point D between points A and B. (To see why, you might review the Triangle Inequality section in Chapter 3 of Practice 3A.) We measure $1\frac{1}{8}$ inches to the left of B, and place D as shown:

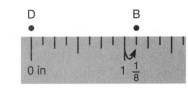

Then, we measure the distance from A to D: **1 inch**.

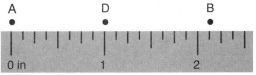

— *or* —

Once we have placed D, the distance from A to D is the difference between the distance from A to B and the distance from B to D. The distance from point A to point B is $2\frac{1}{8}$ inches. The distance from B to D is $1\frac{1}{8}$ inches. So, the distance from A to D is $2\frac{1}{8}-1\frac{1}{8}$ inches.

One inch less than $2\frac{1}{8}$ inches is $1\frac{1}{8}$ inches. Then, $\frac{1}{8}$ of an inch less than $1\frac{1}{8}$ inches is 1 inch. So, the smallest possible distance from A to D is **1 inch**.

164. We measure the distance between the two houses to be $4\frac{2}{8}$ inches. We simplify $\frac{2}{8}$ to $\frac{1}{4}$. In simplest form, the distance from Bobby's house to Carl's is $4\frac{1}{4}$ **inches**.

165. We first measure $1\frac{1}{8}$ inches along the road between the two homes to mark where Bobby stopped.

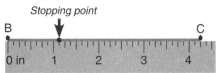

Then, we measure the distance from Bobby's stopping point to Carl's house.

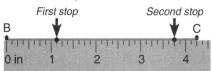

So, when Bobby stops to take a break, he is $3\frac{1}{8}$ **inches** from Carl's house.

— *or* —

The total distance between the houses is $4\frac{1}{4}=4\frac{2}{8}$ inches. Bobby has already walked $1\frac{1}{8}$ inches. After Bobby stops, he needs to walk $1\frac{1}{8}$ inches *less than* the total distance between his house and Carl's.

One inch less than $4\frac{2}{8}$ inches is $3\frac{2}{8}$ inches. Then, $\frac{1}{8}$ of an inch less than $3\frac{2}{8}$ inches is $3\frac{1}{8}$ inches.

So, when Bobby stops to take a break, he is $4\frac{1}{4}-1\frac{1}{8}=4\frac{2}{8}-1\frac{1}{8}=3\frac{1}{8}$ **inches** from Carl's house.

166. We use our drawing from the previous problem and measure $2\frac{5}{8}$ inches from the first to the second stop:

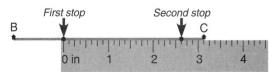

Then we measure the distance from Bobby's house to the second stop.

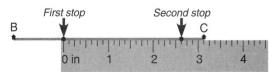

Bobby is $3\frac{6}{8}$ inches from home. We simplify $\frac{6}{8}$ to $\frac{3}{4}$. So at the second stop, Bobby is $3\frac{3}{4}$ **inches** from home.

— *or* —

Bobby walked $1\frac{1}{8}$ inches from home to his first stop and then $2\frac{5}{8}$ inches from his first stop to his second stop. So, he walked a total of $1\frac{1}{8}+2\frac{5}{8}$ inches.

Since $1\frac{1}{8}=1+\frac{1}{8}$ and $2\frac{5}{8}=2+\frac{5}{8}$, we have
$$1\frac{1}{8}+2\frac{5}{8}=1+\frac{1}{8}+2+\frac{5}{8}$$
$$=1+2+\frac{1}{8}+\frac{5}{8}$$

We first add the whole numbers and then the fractions: $1+2$ is 3, and $\frac{1}{8}+\frac{5}{8}$ is $\frac{6}{8}$.

So, $(1+2)+(\frac{1}{8}+\frac{5}{8})=3+\frac{6}{8}=3\frac{6}{8}$.

We simplify $\frac{6}{8}$ to $\frac{3}{4}$.

So, Bobby is $1\frac{1}{8}+2\frac{5}{8}=3\frac{6}{8}=3\frac{3}{4}$ **inches** from home.

167. Bobby walks the distance between the houses twice. We draw one line for Bobby's first trip, then add on the distance of his second trip. The resulting line is the total distance that Bobby walked. The distance from Bobby's to Carl's is $4\frac{1}{4}$ inches. So, we draw a line $4\frac{1}{4}$ inches long.

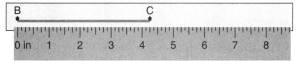

The distance of Bobby's second trip, from Carl's to Bobby's, is also $4\frac{1}{4}$ inches. So, we extend the line $4\frac{1}{4}$ inches, as shown.

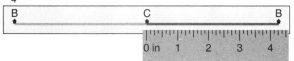

The total length of this line is the total distance that Bobby walks on Tuesday. We measure the length.

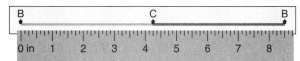

All together, Bobby walks $8\frac{2}{4}$ inches on Tuesday. We simplify $\frac{2}{4}$ to $\frac{1}{2}$, so Bobby walks **$8\frac{1}{2}$ inches** on Tuesday.

— *or* —

The distance between the two houses is $4\frac{1}{4}$ inches, and Bobby walks that distance twice. So, he walks a total of $4\frac{1}{4}+4\frac{1}{4}$ inches. Since $4\frac{1}{4}=4+\frac{1}{4}$, we have

$$4\frac{1}{4}+4\frac{1}{4}=4+\frac{1}{4}+4+\frac{1}{4}$$
$$=4+4+\frac{1}{4}+\frac{1}{4}$$

We first add the whole numbers and then the fractions: $4+4$ is 8, and $\frac{1}{4}+\frac{1}{4}$ is $\frac{2}{4}$.

So, $(4+4)+\left(\frac{1}{4}+\frac{1}{4}\right)=8+\frac{2}{4}=8\frac{2}{4}$.

All together, Bobby walks $8\frac{2}{4}$ inches. We simplify $\frac{2}{4}$ to $\frac{1}{2}$.

So, Bobby walks $4\frac{1}{4}+4\frac{1}{4}=8\frac{2}{4}=\mathbf{8\frac{1}{2}}$ **inches** on Tuesday.

168. To find half of $4\frac{1}{4}$ inches, we cut a piece of paper that is $4\frac{1}{4}$ inches long. Then, we fold the paper in half and measure from one end to the fold, as shown:

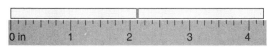

Half the distance from Bobby's house to Carl's house is **$2\frac{1}{8}$ inches**.

— *or* —

To find half of $4\frac{1}{4}=4\frac{2}{8}$ inches, we look for the number that can be added to itself to give $4\frac{2}{8}$. So, we want to solve $\square+\square=4\frac{2}{8}$.

$2+2=4$ and $\frac{1}{8}+\frac{1}{8}=\frac{2}{8}$.

So, $4\frac{2}{8}=2+2+\frac{1}{8}+\frac{1}{8}$
$\qquad\quad=2+\frac{1}{8}+2+\frac{1}{8}$

Since $2+\frac{1}{8}=2\frac{1}{8}$, we have $\boxed{2\frac{1}{8}}+\boxed{2\frac{1}{8}}=4\frac{2}{8}$.

Half of $4\frac{2}{8}$ inches is **$2\frac{1}{8}$ inches**.

---

169. $7\div2=\frac{7}{2}$. Since $\frac{7}{2}$ is one half more than $\frac{6}{2}=3$, we have $7\div2=\frac{7}{2}=\mathbf{3\frac{1}{2}}$.

170. $14\div9=\frac{14}{9}$. Since $\frac{14}{9}$ is five ninths more than $\frac{9}{9}=1$, we have $14\div9=\frac{14}{9}=\mathbf{1\frac{5}{9}}$.

171. $26\div8=\frac{26}{8}$. Since $\frac{26}{8}$ is two eighths more than $\frac{24}{8}=3$, we have $3\frac{2}{8}$. Then, we simplify $\frac{2}{8}$ to $\frac{1}{4}$.
So, in simplest form, $26\div8=\frac{26}{8}=3\frac{2}{8}=\mathbf{3\frac{1}{4}}$.

— *or* —

We simplify $\frac{26}{8}$ to $\frac{13}{4}$. Since $\frac{13}{4}$ is one fourth more than $\frac{12}{4}=3$, we have $26\div8=\frac{26}{8}=\frac{13}{4}=\mathbf{3\frac{1}{4}}$.

172. $86\div10=\frac{86}{10}$. Since $\frac{86}{10}$ is six tenths more than $\frac{80}{10}=8$, we have $86\div10=8\frac{6}{10}$. Then, we simplify $\frac{6}{10}$ to $\frac{3}{5}$.
So, in simplest form, $86\div10=\frac{86}{10}=8\frac{6}{10}=\mathbf{8\frac{3}{5}}$.

— *or* —

We simplify $\frac{86}{10}$ to $\frac{43}{5}$. Since $\frac{43}{5}$ is three fifths more than $\frac{40}{5}=8$, we have $86\div10=\frac{86}{10}=\frac{43}{5}=\mathbf{8\frac{3}{5}}$.

173. $4\div12=\frac{4}{12}$. Since $\frac{4}{12}$ can be simplified to $\frac{1}{3}$, we have $4\div12=\frac{4}{12}=\mathbf{\frac{1}{3}}$.

174. $36\div84=\frac{36}{84}$. We simplify $\frac{36}{84}$:

$$\frac{36}{84}\overset{\div12}{\underset{\div12}{=}}\frac{3}{7}$$

Since $\frac{36}{84}$ can be simplified to $\frac{3}{7}$, we have $36\div84=\frac{36}{84}=\mathbf{\frac{3}{7}}$.

175. To find the side length of a square, we divide its perimeter by its number of sides: $14\div4=\frac{14}{4}$.
Since $\frac{14}{4}$ is two fourths more than $\frac{12}{4}=3$, we have $14\div4=3\frac{2}{4}$. We simplify $\frac{2}{4}$ to $\frac{1}{2}$.
So, $14\div4=\frac{14}{4}=3\frac{2}{4}=3\frac{1}{2}$.

The side length of the square is **$3\frac{1}{2}$ inches**.

— *or* —

We simplify $\frac{14}{4}$ to $\frac{7}{2}$. Then, since $\frac{7}{2}$ is one half more than $\frac{6}{2}=3$, we have $14\div4=\frac{14}{4}=\frac{7}{2}=3\frac{1}{2}$.
The side length of the square is **$3\frac{1}{2}$ inches**.

176. The perimeter of the large triangle is 8 cm. The perimeter of the large triangle is made up of 6 sides of small triangles, so we divide $8\div6$ to find the side length of a small triangle. Since the answer is a measurement, we should write it as a mixed number in simplest form:

$$8\div6=\frac{8}{6}=\frac{4}{3}=1\frac{1}{3}\quad or \quad8\div6=\frac{8}{6}=1\frac{2}{6}=1\frac{1}{3}.$$

So, the side length of a small triangle is **$1\frac{1}{3}$ cm**.

---

177. *Four* sevenths is less than *five* sevenths. So, $\frac{4}{7}$ $\boxed{<}$ $\frac{5}{7}$.

178. We compare $\frac{1}{4}$ to $\frac{3}{8}$ by converting $\frac{1}{4}$ to an equivalent fraction with denominator 8:

$$\frac{1}{4} \overset{\times 2}{\underset{\times 2}{=}} \frac{2}{8}$$

$\frac{2}{8} < \frac{3}{8}$, so $\frac{1}{4}$ $\boxed{<}$ $\frac{3}{8}$.

179. We compare $\frac{3}{4}$ to $\frac{11}{16}$ by converting $\frac{3}{4}$ to an equivalent fraction with denominator 16:

$$\frac{3}{4} \overset{\times 4}{\underset{\times 4}{=}} \frac{12}{16}$$

$\frac{12}{16} > \frac{11}{16}$, so $\frac{3}{4}$ $\boxed{>}$ $\frac{11}{16}$.

180. We compare $\frac{2}{3}$ to $\frac{5}{6}$ by converting $\frac{2}{3}$ to an equivalent fraction with denominator 6:

$$\frac{2}{3} \overset{\times 2}{\underset{\times 2}{=}} \frac{4}{6}$$

$\frac{4}{6} < \frac{5}{6}$, so $\frac{2}{3}$ $\boxed{<}$ $\frac{5}{6}$.

181. We compare $\frac{9}{16}$ to $\frac{5}{8}$ by converting $\frac{5}{8}$ to an equivalent fraction with denominator 16:

$$\frac{5}{8} \overset{\times 2}{\underset{\times 2}{=}} \frac{10}{16}$$

$\frac{9}{16} < \frac{10}{16}$, so $\frac{9}{16}$ $\boxed{<}$ $\frac{5}{8}$.

182. We compare $\frac{10}{21}$ to $\frac{3}{7}$ by converting $\frac{3}{7}$ to an equivalent fraction with denominator 21:

$$\frac{3}{7} \overset{\times 3}{\underset{\times 3}{=}} \frac{9}{21}$$

$\frac{10}{21} > \frac{9}{21}$, so $\frac{10}{21}$ $\boxed{>}$ $\frac{3}{7}$.

183. Eighths are larger than elevenths. So, $\frac{5}{8}$ $\boxed{>}$ $\frac{5}{11}$.

184. We compare $\frac{2}{7}$ to $\frac{4}{13}$ by converting $\frac{2}{7}$ to an equivalent fraction with numerator 4:

$$\frac{2}{7} \overset{\times 2}{\underset{\times 2}{=}} \frac{4}{14}$$

Fourteenths are smaller than thirteenths, so $\frac{4}{14} < \frac{4}{13}$. Therefore, $\frac{2}{7}$ $\boxed{<}$ $\frac{4}{13}$.

185. We compare $\frac{3}{5}$ to $\frac{6}{11}$ by converting $\frac{3}{5}$ to an equivalent fraction with numerator 6:

$$\frac{3}{5} \overset{\times 2}{\underset{\times 2}{=}} \frac{6}{10}$$

Tenths are larger than elevenths, so $\frac{6}{10} > \frac{6}{11}$. Therefore, $\frac{3}{5}$ $\boxed{>}$ $\frac{6}{11}$.

186. We compare $\frac{4}{9}$ to $\frac{20}{43}$ by converting $\frac{4}{9}$ to an equivalent fraction with numerator 20:

$$\frac{4}{9} \overset{\times 5}{\underset{\times 5}{=}} \frac{20}{45}$$

Forty-fifths are smaller than forty-thirds, so $\frac{20}{45} < \frac{20}{43}$. Therefore, $\frac{4}{9}$ $\boxed{<}$ $\frac{20}{43}$.

187. We compare $\frac{9}{31}$ to $\frac{3}{11}$ by converting $\frac{3}{11}$ to an equivalent fraction with numerator 9:

$$\frac{3}{11} \overset{\times 3}{\underset{\times 3}{=}} \frac{9}{33}$$

Thirty-firsts are larger than thirty-thirds, so $\frac{9}{31} > \frac{9}{33}$. Therefore, $\frac{9}{31}$ $\boxed{>}$ $\frac{3}{11}$.

188. We compare $\frac{24}{43}$ to $\frac{4}{7}$ by converting $\frac{4}{7}$ to an equivalent fraction with numerator 24:

$$\frac{4}{7} \overset{\times 6}{\underset{\times 6}{=}} \frac{24}{42}$$

Forty-thirds are smaller than forty-seconds, so $\frac{24}{43} < \frac{24}{42}$. Therefore, $\frac{24}{43}$ $\boxed{<}$ $\frac{4}{7}$.

189. We write fractions equivalent to $\frac{1}{2}$ that have denominators of 6, 8, 10, and 12:

$$\frac{1}{2} = \frac{3}{6} = \frac{4}{8} = \frac{5}{10} = \frac{6}{12}$$

Then, we use these equivalent fractions to compare fractions with the same denominators.

$$\frac{2}{6} < \frac{3}{6} < \frac{4}{6}$$
$$\frac{3}{8} < \frac{4}{8} < \frac{5}{8}$$
$$\frac{4}{10} < \frac{5}{10} < \frac{6}{10}$$
$$\frac{5}{12} < \frac{6}{12} < \frac{7}{12}$$

So, $\frac{4}{6}$, $\frac{5}{8}$, $\frac{6}{10}$, and $\frac{7}{12}$ are the fractions on the list that are greater than $\frac{1}{2}$. The other four are less than $\frac{1}{2}$.

— *or* —

If the numerator of a fraction is half its denominator, then the fraction equals $\frac{1}{2}$. If the numerator of a fraction is more than half its denominator, then the fraction is greater than $\frac{1}{2}$. So, we look for any fraction whose numerator is more than half its denominator.

Since 4 is more than half of 6, and 5 is more than half of 8, and 6 is more than half of 10, and 7 is more than half of 12, we see $\frac{4}{6}$, $\frac{5}{8}$, $\frac{6}{10}$, and $\frac{7}{12}$ are all greater than $\frac{1}{2}$. In each of the four other fractions, the numerator is less than half the denominator, so those fractions are less than $\frac{1}{2}$.

$\frac{2}{6}$ $\boxed{\frac{4}{6}}$ $\frac{3}{8}$ $\boxed{\frac{5}{8}}$ $\frac{4}{10}$ $\boxed{\frac{6}{10}}$ $\frac{5}{12}$ $\boxed{\frac{7}{12}}$

190. We write fractions equivalent to $\frac{1}{2}$ that have numerators of 3, 4, 5, and 6:

$$\frac{1}{2}=\frac{3}{6}=\frac{4}{8}=\frac{5}{10}=\frac{6}{12}$$

Then, we use these equivalent fractions to compare fractions with the same numerators:

$$\frac{3}{7}<\frac{3}{6}<\frac{3}{5}$$
$$\frac{4}{9}<\frac{4}{8}<\frac{4}{7}$$
$$\frac{5}{11}<\frac{5}{10}<\frac{5}{9}$$
$$\frac{6}{13}<\frac{6}{12}<\frac{6}{11}$$

So, $\frac{3}{7}$, $\frac{4}{9}$, $\frac{5}{11}$, and $\frac{6}{13}$ are the fractions on the list that are less than $\frac{1}{2}$. The other four are greater than $\frac{1}{2}$.

— *or* —

If the numerator of a fraction is half its denominator, then the fraction equals $\frac{1}{2}$. If the numerator of a fraction is less than half its denominator, then the fraction is less than $\frac{1}{2}$. So, we look for any fraction whose numerator is less than half its denominator.

Since 3 is less than half of 7, and 4 is less than half of 9, and 5 is less than half of 11, and 6 is less than half of 13, $\frac{3}{7}$, $\frac{4}{9}$, $\frac{5}{11}$, and $\frac{6}{13}$ are all less than $\frac{1}{2}$. In each of the four other fractions, the numerator is greater than half the denominator, so those fractions are greater than $\frac{1}{2}$.

$$\frac{3}{5} \quad \boxed{\frac{3}{7}} \quad \frac{4}{7} \quad \boxed{\frac{4}{9}} \quad \frac{5}{9} \quad \boxed{\frac{5}{11}} \quad \frac{6}{11} \quad \boxed{\frac{6}{13}}$$

191. To compare $\frac{4}{9}$, $\frac{2}{3}$, and $\frac{8}{15}$, we write the fractions with a common numerator or denominator. Eight is a multiple of both 4 and 2, so we convert both $\frac{4}{9}$ and $\frac{2}{3}$ to equivalent fractions with numerator 8:

$$\frac{4}{9} \overset{\times 2}{=} \frac{8}{18} \text{ and } \frac{2}{3} \overset{\times 4}{=} \frac{8}{12}$$

Fifteenths are smaller than twelfths, so $\frac{8}{15}<\frac{8}{12}$.

Eighteenths are smaller than fifteenths, so $\frac{8}{18}<\frac{8}{15}$. So, we have $\frac{8}{18}<\frac{8}{15}<\frac{8}{12}$. In simplest form, we order the fractions from least to greatest: $\frac{4}{9}$, $\frac{8}{15}$, and $\frac{2}{3}$.

192. The numerators of the fractions are all the same, so we only need to compare the denominators of the fractions. Since sixths are larger than sevenths, we have $\frac{5}{7}<\frac{5}{6}$. Since sevenths are larger than eighths, we have $\frac{5}{8}<\frac{5}{7}$. So, we place the numbers as follows to make a true statement: $\frac{5}{8}<\frac{5}{7}<\frac{5}{6}$.

193. All fractions must be in simplest form.

8 cannot be placed below 4 or 6 to create a fraction in simplest form, so 8 must be placed below 5: $\frac{4}{}<\frac{5}{8}<\frac{6}{}$.

Similarly, 9 cannot be placed below 6 to create a fraction in simplest form, so 9 must be placed below the 4: $\frac{4}{9}<\frac{5}{8}<\frac{6}{}$. This leaves us to place the 7 below the 6:

$$\frac{4}{9}<\frac{5}{8}<\frac{6}{7}.$$

We check our answers by comparing each pair of fractions: $\frac{4}{9}<\frac{5}{8}$ because $\frac{4}{9}<\frac{5}{9}<\frac{5}{8}$. $\frac{5}{8}<\frac{6}{7}$ because $\frac{5}{8}<\frac{6}{8}<\frac{6}{7}$. So, $\frac{4}{9}<\frac{5}{8}<\frac{6}{7}$. ✓

194. The fractions must be less than 1. So, we cannot place 7 or 11 above the 7. We can only place the 3 above the 7. Then, we have $\frac{3}{7}<\frac{3}{}<\frac{5}{}$.

We also cannot place the 7 below the 3, since $\frac{3}{7}$ is *equal to* (not less than) $\frac{3}{7}$. So, the 7 must be placed below the 5, giving us $\frac{3}{7}<\frac{3}{}<\frac{5}{7}$.

This leaves us to place the 11 below the 3: $\frac{3}{11}<\frac{3}{7}<\frac{5}{7}$.

We check our answers: the first two fractions have the same numerator and the second two fractions have the same denominator. Since elevenths are smaller than sevenths, $\frac{3}{11}<\frac{3}{7}$. Then, $\frac{3}{7}<\frac{5}{7}$. So, $\frac{3}{11}<\frac{3}{7}<\frac{5}{7}$. ✓

195. All fractions must be in simplest form. So, only the 3 or 5 can be placed below the 2.

If we place the 5 below the 2, we get $\frac{}{7}<\frac{2}{5}<\frac{}{}$. Then, because the fractions must be in simplest form, we cannot place the 6 with 3 or 4. This leaves us to place the 6 above the 7. However, $\frac{6}{7}$ is not less than $\frac{2}{5}$. $\left(\frac{2}{5}=\frac{6}{15}, \text{ which is less than } \frac{6}{7}.\right)$ So, we cannot place the 5 below the 2. ✗

If we place the 3 below the 2, we get $\frac{}{7}<\frac{2}{3}<\frac{}{}$. Since $\frac{5}{7}$ and $\frac{6}{7}$ are both greater than $\frac{2}{3}$, we can only place the 4 above the 7. $\left(\frac{5}{7}=\frac{10}{14}\text{ is greater than }\frac{2}{3}=\frac{10}{15}.\right)$ So, we have $\frac{4}{7}<\frac{2}{3}<\frac{}{}$. This leaves us to place 5 and 6 as the numerator and denominator of the final fraction. Since each fraction must be less than 1, the numerator must be 5 and the denominator must be 6. This gives us $\frac{4}{7}<\frac{2}{3}<\frac{5}{6}$.

We check this statement by converting $\frac{2}{3}$ to $\frac{4}{6}$ and comparing pairs of fractions: $\frac{4}{7}<\frac{4}{6}<\frac{5}{6}$. $\frac{4}{7}<\frac{4}{6}$, so $\frac{4}{7}<\frac{2}{3}$. Also, $\frac{4}{6}<\frac{5}{6}$, so $\frac{2}{3}<\frac{5}{6}$. Therefore, we have $\frac{4}{7}<\frac{2}{3}<\frac{5}{6}$. ✓

196. We can compare the numbers $\frac{1}{2}$ and $\frac{2}{3}$, and we know that $\frac{2}{3}>\frac{2}{4}=\frac{1}{2}$. However, this problem is about fractions of two *different* wholes. We cannot tell who ate more because we do not know the size of the cake Alex ate or the size of the pie Grogg ate.

For example, Alex might have eaten $\frac{1}{2}$ of a huge cake and Grogg might have eaten $\frac{2}{3}$ of a tiny pie. Or, Alex might have eaten $\frac{1}{2}$ of a tiny cake and Grogg might have eaten $\frac{2}{3}$ of a huge pie.

We need more information to know who ate more.

Compare this to a problem with whole numbers: *Alex ate 2 cakes. Grogg ate 3 pies. Who ate more?* We know that $3>2$, but the cakes might have been bigger than the pies. Without more information, we cannot know who ate more!

197. We see from the number line that only one marked point is less than $\frac{1}{2}$. Since $\frac{3}{7}$ is the only given fraction whose numerator is less than half its denominator, $\frac{3}{7}$ is the marked point that is less than $\frac{1}{2}$:

Then, we compare the three remaining fractions to each other: $\frac{3}{4}$, $\frac{6}{7}$, and $\frac{5}{8}$. We convert $\frac{3}{4}$ to $\frac{6}{8}$, then compare it to the other two numbers.

First, we know $\frac{5}{8} < \frac{6}{8}$, so $\frac{5}{8} < \frac{3}{4}$. Then, $\frac{6}{8} < \frac{6}{7}$, so $\frac{3}{4} < \frac{6}{7}$.

Therefore, we have $\frac{5}{8} < \frac{3}{4} < \frac{6}{7}$.

We label the remaining points in that order:

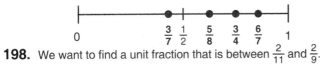

198. We want to find a unit fraction that is between $\frac{2}{11}$ and $\frac{2}{9}$.

Since tenths are larger than elevenths, and ninths are larger than tenths, we have $\frac{2}{11} < \frac{2}{10} < \frac{2}{9}$.

We simplify $\frac{2}{10}$ to a unit fraction: $\frac{2}{10} = \frac{1}{5}$. Then, writing each fraction in simplest form, we get $\frac{2}{11} < \frac{1}{5} < \frac{2}{9}$.

Therefore, $y = \mathbf{5}$.

199. We convert these two fractions to equivalent fractions with equal numerators to compare them:

$$\frac{2}{5} \xrightarrow{\times 7} \frac{14}{35} \quad \text{and} \quad \frac{7}{21} \xrightarrow{\times 2} \frac{14}{42}$$

Since $\frac{14}{35} > \frac{14}{42}$, we have $\frac{2}{5} \enclose{circle}{>} \frac{7}{21}$.

— *or* —

We convert these two fractions to equivalent fractions with equal denominators to compare them:

$$\frac{2}{5} \xrightarrow{\times 21} \frac{42}{105} \quad \text{and} \quad \frac{7}{21} \xrightarrow{\times 5} \frac{35}{105}$$

Since $\frac{42}{105} > \frac{35}{105}$, we have $\frac{2}{5} \enclose{circle}{>} \frac{7}{21}$.

— *or* —

Notice that $\frac{7}{21}$ is not in simplest form. We simplify $\frac{7}{21}$ to $\frac{1}{3}$. Then, to compare this to $\frac{2}{5}$, we convert $\frac{1}{3}$ to an equivalent fraction with numerator 2.

$$\frac{1}{3} \xrightarrow{\times 2} \frac{2}{6}$$

Since fifths are larger than sixths, $\frac{2}{5} > \frac{2}{6}$. So, $\frac{2}{5} > \frac{1}{3}$.

Therefore, $\frac{2}{5} \enclose{circle}{>} \frac{7}{21}$.

200. We convert these two fractions to equivalent fractions with equal denominators to compare them.

$$\frac{37}{99} \xrightarrow{\times 200} \frac{7,400}{19,800} \quad \text{and} \quad \frac{73}{200} \xrightarrow{\times 99} \frac{7,227}{19,800}$$

Since $\frac{7,400}{19,800} > \frac{7,227}{19,800}$, we have $\frac{37}{99} \enclose{circle}{>} \frac{73}{200}$.

— *or* —

We look at the denominators of both fractions and notice that 99 is about half of 200. So, we convert $\frac{37}{99}$ to an equivalent fraction by multiplying each of the numerator and denominator by 2.

$$\frac{37}{99} \xrightarrow{\times 2} \frac{74}{198}$$

Now, we compare $\frac{74}{198}$ to $\frac{73}{200}$.

198ths are larger than 200ths. There are more 198ths in $\frac{74}{198}$ than 200ths in $\frac{73}{200}$. Since there are more 198ths and 198ths are larger than 200ths, we find $\frac{74}{198} > \frac{73}{200}$.

Therefore, $\frac{37}{99} \enclose{circle}{>} \frac{73}{200}$.

— *or* —

Since 198ths are larger than 200ths, we have $\frac{74}{198} > \frac{74}{200}$. Then, since 73 is less than 74, we have $\frac{74}{200} > \frac{73}{200}$. So, $\frac{74}{198} > \frac{74}{200} > \frac{73}{200}$. Therefore, $\frac{74}{198} > \frac{73}{200}$ and $\frac{37}{99} \enclose{circle}{>} \frac{73}{200}$.

To compare two fractions, it can sometimes be useful to convert one or both fractions into equivalent fractions with close denominators. However, this approach does not always work. For example, this approach does not help us compare $\frac{36}{99}$ to $\frac{73}{200}$. Converting $\frac{36}{99}$ as above, we get $\frac{36}{99} = \frac{72}{198}$. There are fewer 198ths in $\frac{72}{198}$ than 200ths in $\frac{73}{200}$, but 198ths are larger than 200ths. So, we cannot tell which is greater without more computation.

201. We convert the known fractions into fractions with the same numerator to look for other fractions between them:

$$\frac{2}{7} \xrightarrow{\times 3} \frac{6}{21} \quad \text{and} \quad \frac{3}{8} \xrightarrow{\times 2} \frac{6}{16}$$

A fraction with numerator 6 is equivalent to a unit fraction if the denominator is a multiple of 6.

(For example, $\frac{6}{12} = \frac{1}{2}$ and $\frac{6}{42} = \frac{1}{7}$.)

The only fraction between $\frac{6}{21}$ and $\frac{6}{16}$ that has numerator 6 and a denominator that is a multiple of 6 is $\frac{6}{18}$.

So, we have $\frac{6}{21} < \frac{6}{18} < \frac{6}{16}$. Writing each fraction in simplest form, we have $\frac{2}{7} < \frac{1}{3} < \frac{3}{8}$. Therefore, $z = \mathbf{3}$.

— *or* —

We can compare $\frac{2}{7}$ and $\frac{3}{8}$ to nearby unit fractions.

$\frac{2}{7}$ is greater than $\frac{2}{8} = \frac{1}{4}$ but less than $\frac{2}{6} = \frac{1}{3}$.

So, $\frac{1}{4} < \frac{2}{7} < \frac{1}{3}$.

$\frac{3}{8}$ is greater than $\frac{3}{9} = \frac{1}{3}$ but less than $\frac{3}{6} = \frac{1}{2}$.

So, $\frac{1}{3} < \frac{3}{8} < \frac{1}{2}$.

Putting this all together, we have:

$$\frac{1}{4} < \frac{2}{7} < \frac{1}{3} < \frac{3}{8} < \frac{1}{2}.$$

So, $\frac{1}{3}$ is greater than $\frac{2}{7}$, but less than $\frac{3}{8}$. Therefore, $z = \mathbf{3}$.

202. There are six possible ways to arrange the digits 7, 8, and 9 to create a fraction that is less than 1. We organize these by numerator:

$$\frac{7}{89} \qquad \frac{7}{98} \qquad \frac{8}{79} \qquad \frac{8}{97} \qquad \frac{9}{78} \qquad \frac{9}{87}$$

To compare these fractions to $\frac{1}{10}$ and $\frac{1}{9}$, we convert $\frac{1}{10}$ and $\frac{1}{9}$ to equivalent fractions with numerators 7, 8 and 9.

$$\frac{1}{10} = \frac{7}{70} = \frac{8}{80} = \frac{9}{90} \text{ and } \frac{1}{9} = \frac{7}{63} = \frac{8}{72} = \frac{9}{81}.$$

Lizzie's number is *between* $\frac{1}{10}$ and $\frac{1}{9}$.

$\frac{7}{89}$ is less than both $\frac{7}{70}$ and $\frac{7}{63}$. ✗

$\frac{7}{98}$ is less than both $\frac{7}{70}$ and $\frac{7}{63}$. ✗

$\frac{8}{79}$ is greater than $\frac{8}{80}$ and less than $\frac{8}{72}$. ✓

$\frac{8}{97}$ is less than both $\frac{8}{80}$ and $\frac{8}{72}$. ✗

$\frac{9}{78}$ is greater than both $\frac{9}{90}$ and $\frac{9}{81}$. ✗

$\frac{9}{87}$ is greater than $\frac{9}{90}$ and less than $\frac{9}{81}$. ✓

$\frac{8}{79}$ and $\frac{9}{87}$ are the only two numbers between $\frac{1}{10}$ and $\frac{1}{9}$.
So, Lizzie could have written **$\frac{8}{79}$** or **$\frac{9}{87}$**.

FRACTIONS
Splitting Shapes
44-45

You may have split the shapes differently than the solutions shown. However, be sure that each piece is the same size!

You may have shaded different pieces than those shown. However, be sure that you have shaded the correct number of pieces!

203. We split the shape into 2 congruent L-shaped pieces and shade $\frac{1}{2}$ of the shape.

204. We split the shape into 5 congruent squares and shade $\frac{2}{5}$ of the shape.

— *or* —

We split the shape into 10 congruent right triangles and shade $\frac{4}{10} = \frac{2}{5}$ of the shape.

205. We split the shape into 8 congruent isosceles triangles and shade $\frac{3}{8}$ of the shape.

206. We split the shape into 11 congruent isosceles triangles and shade $\frac{4}{11}$ of the shape.

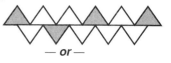

— *or* —

We split the shape into 22 congruent right triangles and shade $\frac{8}{22} = \frac{4}{11}$ of the shape.

207. We split the shape into 6 congruent pentagons and shade $\frac{5}{6}$ of the shape.

208. We split the shape into 9 congruent right triangles and shade $\frac{4}{9}$ of the shape.

209. We split the shape into 10 congruent right triangles and shade $\frac{7}{10}$ of the shape.

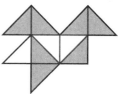

210. We split the shape into 4 congruent pieces and shade $\frac{3}{4}$ of the shape.

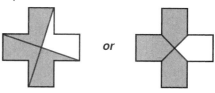   *or*

211. We split the shape into 3 congruent quadrilaterals and shade $\frac{1}{3}$ of the shape.

212. We split the shape into 7 congruent right triangles and shade $\frac{5}{7}$ of the shape.

ESTIMATION
When to Estimate page 47

1. It is nearly impossible to count the exact number of hairs on your head, so an **estimate** is most appropriate.

2. We want to make sure that every student who should be on the bus is there. Having even one fewer or one more student than we expect means that someone is missing or someone is on the bus who shouldn't be! So, it is important to know the **exact** number of students on the bus ride back from a field trip.

3. The exact distance from Earth to the Moon varies throughout the year and is very difficult to calculate. So, an **estimate** is most appropriate. (The exact distance varies from about 357,000 miles to about 406,000 miles.)

4. A basketball game can be won by just one point. So, it is important to know the **exact** number of points scored by each team in a basketball game.

5. A carpenter needs to know the **exact** width of a door that he or she will install. If a door is even a fraction of an inch too large or small, it may not fit properly.

6. The exact number of gallons that flow down a huge river each minute is nearly impossible to compute and is constantly changing. So, an **estimate** is most appropriate.

7. It is important to know the **exact** height of the bar in a high-jump event. The winner may jump only a fraction of an inch higher than the second-place high jumper.

8. The total number of apples in an orchard is very difficult to compute, and knowing the exact number is probably not necessary. An **estimate** is most appropriate.

ESTIMATION
Rounding Whole Numbers 48-50

9. Since 62 is closer to 60 than to 70, we round 62 down to **60**.

10. Since 486 is closer to 490 than to 480, we round 486 up to **490**.

11. Since 101 is closer to 100 than to 110, we round 101 down to **100**.

12. Since 1,395 is exactly between 1,390 and 1,400, it is not closer to either number. Numbers in the middle are rounded up. So, 1,395 rounds up to **1,400**.

13. Since 1,283 is closer to 1,300 than to 1,200, we round 1,283 up to **1,300**.

14. Since 5,135 is closer to 5,100 than to 5,200, we round 5,135 down to **5,100**.

15. Since 44,445 is closer to 44,400 than to 44,500, we round 44,445 down to **44,400**.

16. Since 199,999 is closer to 200,000 than to 199,900, we round 199,999 up to **200,000**.

17. Since 7,890 is closer to 8,000 than to 7,000, we round 7,890 up to **8,000**.

18. Since 45,678 is closer to 46,000 than to 45,000, we round 45,678 up to **46,000**.

19. Since 2,000,499 is closer to 2,000,000 than to 2,001,000, we round 2,000,499 down to **2,000,000**.

20. Since 99,615 is closer to 100,000 than to 99,000, we round 99,615 up to **100,000**.

21. All of the whole numbers from 250 to 349 round to 300 when rounded to the nearest hundred. So, the smallest possible value of Bill's number is **250**.

22. All of the whole numbers from 5,500 to 6,499 round to 6,000 when rounded to the nearest thousand. So, the largest whole number that rounds to 6,000 when rounded to the nearest thousand is **6,499**.

23. All of the whole numbers from 45 to 54 round to 50 when rounded to the nearest ten. All together, there are **10** whole numbers that round to 50 when rounded to the nearest ten: 45, 46, 47, 48, 49, 50, 51, 52, 53, and 54.

24. Kim rounds 777 to the nearest ten and gets 780. Jim rounds 777 to the nearest hundred and gets 800. Jim's estimate is larger than Kim's estimate by $800 - 780 = $ **20**.

25. Every 2-digit number is less than 100. So, Kyle must have rounded his favorite number *up* to 100. Kyle's estimate is 18 more than his number, so Kyle's favorite two-digit number is $100 - 18 = $ **82**.

26. Kelly's number must be 36 less than a multiple of 100. We look for a number between 582 and 761 that is 36 less than a multiple of 100.

$600 - 36 = 564$ $700 - 36 = 664$ $800 - 36 = 764$

Only 664 is between 582 and 761. So, Kelly's favorite number is **664**.

27. All of the whole numbers from 7,995 to 8,004 round to 8,000 when rounded to the nearest ten.
All of the whole numbers from 7,950 to 8,049 round to 8,000 when rounded to the nearest hundred.
All of the whole numbers from 7,500 to 8,499 round to 8,000 when rounded to the nearest thousand.
The smallest number that is in all three groups is **7,995**.

28. All of the whole numbers from 350 to 449 round to 400 when rounded to the nearest hundred.
All of the whole numbers from 445 to 454 round to 450 when rounded to the nearest ten.
The only numbers that appear in both groups are 445, 446, 447, 448, and 449.
All together, there are **5** numbers.

29. We work backwards, finding the smallest possible value of each little monster's number.

Alex rounds Lizzie's number to the nearest thousand and gets 9,000. All of the whole numbers from 8,500 to 9,499 round to 9,000 when rounded to the nearest thousand. So, the smallest possibility for Lizzie's number is 8,500.

Lizzie rounds Grogg's number to the nearest hundred and gets 8,500. All of the whole numbers from 8,450 to 8,549 round to 8,500 when rounded to the nearest hundred. So, the smallest possibility for Grogg's number is 8,450.

Grogg rounds Winnie's number to the nearest ten and gets 8,450. All of the whole numbers from 8,445 to 8,454 round to 8,450 when rounded to the nearest ten. So, the smallest possibility for Winnie's number is **8,445**.

We check our work: Winnie chooses 8,445. Grogg rounds 8,445 to the nearest ten and gets 8,450. Lizzie then rounds 8,450 to the nearest hundred and gets 8,500. Alex then rounds 8,500 the nearest thousand and gets 9,000. ✓

Note that when rounded to the nearest thousand, 8,445 rounds to 8,000.

ESTIMATION
Rounding Fractions 51

30. The fractional part of $12\frac{1}{3}$ is $\frac{1}{3}$. Since $\frac{1}{3}$ is less than $\frac{1}{2}$, we round $12\frac{1}{3}$ down to **12**.

31. The fractional part of $15\frac{3}{4}$ is $\frac{3}{4}$. Since 3 is more than half of 4, we know that $\frac{3}{4}$ is greater than $\frac{1}{2}$. So, we round $15\frac{3}{4}$ up to **16**.

32. The fractional part of $23\frac{5}{9}$ is $\frac{5}{9}$. Since 5 is more than half of 9, we know that $\frac{5}{9}$ is more than $\frac{1}{2}$. So, we round $23\frac{5}{9}$ up to **24**.

33. First, we write $\frac{9}{7}$ as a mixed number: $\frac{9}{7}=1\frac{2}{7}$. The fractional part of $1\frac{2}{7}$ is $\frac{2}{7}$. Since 2 is less than half of 7, we know that $\frac{2}{7}$ is less than $\frac{1}{2}$. So, we round $1\frac{2}{7}$ down to **1**.

34. First, we write $\frac{19}{5}$ as a mixed number: $\frac{19}{5}=3\frac{4}{5}$. The fractional part of $3\frac{4}{5}$ is $\frac{4}{5}$. Since 4 is more than half of 5, we know that $\frac{4}{5}$ is more than $\frac{1}{2}$. So, we round $3\frac{4}{5}$ up to **4**.

35. First, we write $\frac{35}{8}$ as a mixed number: $\frac{35}{8}=4\frac{3}{8}$. The fractional part of $4\frac{3}{8}$ is $\frac{3}{8}$. Since 3 is less than half of 8, we know that $\frac{3}{8}$ is less than $\frac{1}{2}$. So, we round $4\frac{3}{8}$ down to **4**.

36. The fractional part of $29\frac{50}{99}$ is $\frac{50}{99}$. Since 50 is more than half of 99, we know that $\frac{50}{99}$ is more than $\frac{1}{2}$. So, we round $29\frac{50}{99}$ up to **30**.

37. First, we write $\frac{87}{11}$ as a mixed number: $\frac{87}{11}=7\frac{10}{11}$. The fractional part of $7\frac{10}{11}$ is $\frac{10}{11}$. Since 10 is more than half of 11, we know that $\frac{10}{11}$ is more than $\frac{1}{2}$. So, we round $7\frac{10}{11}$ up to **8**.

— *or* —

We notice that $\frac{87}{11}$ is ten elevenths more than $\frac{77}{11}=7$ and only one eleventh less than $\frac{88}{11}=8$. Since $\frac{87}{11}$ is closer to 8 than to 7, we round $\frac{87}{11}$ up to **8**.

ESTIMATION
Computing Estimates 52-54

38. Since 5,796 is about 6,000 and 9,359 is about 9,000, we estimate that 5,796+9,359 is about 6,000+9,000 = 15,000.
Our estimate suggests that 5,796+9,358 is closer to **15,000** than to 150,000.

In fact, 5,796+9,359 = 15,155.

39. Since 1,123 is about 1,000 and 47 is about 50, we estimate that 1,123×47 is about 1,000×50 = 50,000. Our estimate suggests that 1,123×47 is closer to **50,000** than to 5,000.

In fact, 1,123×47 = 52,781.

40. Since $8\frac{3}{4}$ is about 9 and $13\frac{1}{3}$ is about 13, we estimate that $8\frac{3}{4}+13\frac{1}{3}$ is about 9+13 = 22. Since 22 is closer to 20 than to 30, our estimate suggests that $8\frac{3}{4}+13\frac{1}{3}$ is closer to **20** than to 30.

In fact, $8\frac{3}{4}+13\frac{1}{3}=22\frac{1}{12}$.

41. Since 72,345 is about 70,000 and 68,922 is about 70,000, we estimate that 72,345+68,922 is about 70,000+70,000 = 140,000.
Our estimate has 6 digits. It is much larger than the largest 5-digit number (99,999) and much smaller than the smallest 7-digit number (1,000,000). So, our estimate suggests that the sum 72,345+68,922 also has **6** digits.

In fact, 72,345+68,922 = 141,267.

42. Since 37 is about 40 and 27 is about 30, we estimate that 37×27 is about 40×30 = 1,200. Of the four choices, **999** is the only number that is close enough to our estimate to be correct.

555　　(999)　　6,789　　9,999

43. Grogg rounds 89 to 100 and 203 to 200. So, Grogg's estimate of 89×203 is 100×200 = 20,000.
Alex rounds 89 to 90 and 203 to 200.
Alex's estimate of 89×203 is 90×200 = 18,000.
The difference between these two estimates is 20,000−18,000 = 2,000.
So, Alex's estimate is **2,000** more than Grogg's.

The actual value of 89×203 is 18,067.

44. Since 231 is about 200 and 533 is about 500, we estimate that 231×533 is about 200×500 = 100,000.

Of the four choices, **123,123** is the only number close enough to our estimate to be correct.

1,234　　12,321　　(123,123)　　1,234,567

45. Since $24\frac{1}{3}$ is about 24 and $7\frac{8}{9}$ is about 8, we estimate that $24\frac{1}{3}\div7\frac{8}{9}$ is about 24÷8 = **3**.

In fact, $24\frac{1}{3}\div7\frac{8}{9}=3\frac{6}{71}$.

46. When Adam estimates, he rounds 6 to 10.
When Jon estimates, he rounds 704 to 700.
The amount we add to get from 6 to 10 is the same as the amount we subtract to get from 704 to 700. In both cases, the rounded number differs from the actual number by 4.

However, since 4 is almost as big as 6, increasing 6 by 4 gives us a number that is almost double the original. So, when Adam estimates 6×704 by computing 10×704, his result is almost double the actual value!

But, 4 is very small compared to 704. For example, if you have a jar that holds 704 pennies and someone removes 4, you probably wouldn't even notice. Rounding 704 to 700 gives Jon a very good estimate of the product.

When estimating the product, it is much better to round 704 to 700 than to round 6 to 10. So, Jon's estimate of $6 \times 700 = 4,200$ is much closer to the actual value of 6×704 than Adam's estimate of $10 \times 704 = 7,040$.

Be careful when estimating a product! If your rounded value for a number is almost double the original number, then you probably won't make a very good estimate.

47. When Globb rounds each number to the nearest hundred, he rounds 49 to 0, and 499 to 500, and 4,999 to 5,000. So, **his estimate of the product of these three numbers is $0 \times 500 \times 5,000 = 0$.**
The product $49 \times 499 \times 4,999$ is obviously much greater than 0. So, **Globb's estimate is *not* a good estimate.**
Since any number multiplied by 0 will always equal 0, we avoid rounding a number in a product to 0. Instead, **Globb should have rounded 49 to the nearest ten.**
Since $49 \approx 50$, $499 \approx 500$, and $4,999 \approx 5,000$, a good estimate of $49 \times 499 \times 4,999$ is $50 \times 500 \times 5,000 = 125,000,000$.
The actual value of $49 \times 499 \times 4,999$ is $122,230,549$.

48. ☑ $79 \times 107 \approx 80 \times 100 = 8,000$. Since 8,000 is close to 8,453, the computation is reasonable.

49. ☒ $1,297 + 680 \approx 1,300 + 700 = 2,000$. Since 2,000 is nowhere near 8,097, the computation is not reasonable.

50. ☒ $67 \times 1,008 \approx 67 \times 1,000 = 67,000$. Since 67,000 is nowhere near 7,236, the computation is not reasonable.

51. ☑ $532 + 118 \approx 500 + 100 = 600$. Since 600 is close to 650, the computation is reasonable.

52. ☒ $89 + 778 \approx 90 + 800 = 890$. Since 890 is nowhere near 1,668, the computation is not reasonable.

53. ☑ $603 \times 90 \approx 600 \times 90 = 54,000$. Since 54,000 is close to 54,270, the computation is reasonable.

54. ☒ $7 \times 986 \approx 7 \times 1,000 = 7,000$. Since 7,000 is nowhere near 69,020, the computation is not reasonable.

55. ☒ $78 \times 215 \approx 80 \times 200 = 16,000$. Since 16,000 is nowhere near 1,677, the computation is not reasonable.

56. ☑ $977 + 14,040 \approx 1,000 + 14,000 = 15,000$. Since 15,000 is close to 15,017, the computation is reasonable.

57. ☑ $9 \times 11 \times 8 \approx 10 \times 10 \times 10 = 1,000$. Since 1,000 is close to 792, the computation is reasonable.

58. ☑ $866 + 1,671 \approx 900 + 1,700 = 2,600$. Since 2,600 is close to 2,537, the computation is reasonable.

59. ☒ $7 \times (89 + 23) \approx 7 \times (90 + 20) = 7 \times 110 = 770$. Since 770 is nowhere near 7,784, the computation is not reasonable.

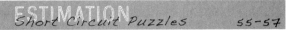

ESTIMATION *Short Circuit Puzzles* 55-57

For the following problems, you may have estimated differently to match the same pairs of expressions.

60. We first estimate the sums on the left:
$96 + 78 \approx 100 + 80 = 180$.
$58 + 39 \approx 60 + 40 = 100$.
$88 + 49 \approx 90 + 50 = 140$.

Then, we compare these estimates to the values on the right. Since 180 is closest to 174, we pair $96 + 78$ and 174. Since 100 is closest to 97, we pair $58 + 39$ and 97. Since 140 is closest to 137, we pair $88 + 49$ and 137.

We connect these pairs without crossing wires as shown:

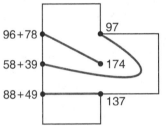

61. We estimate the products on the left:
$6 \times 23 \approx 6 \times 20 = 120$.
$28 \times 86 \approx 30 \times 90 = 2,700$.
$104 \times 12 \approx 100 \times 12 = 1,200$.

Then, we compare these estimates to the values on the right. Since 120 is closest to 138, we pair 6×23 and 138. Since 2,700 is closest to 2,408, we pair 28×86 and 2,408. Since 1,200 is closest to 1,248, we pair 104×12 and 1,248.

We connect these pairs without crossing wires as shown:

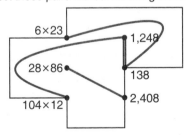

62. We first estimate the sums on the left:
$$186+292 \approx 200+300 = 500.$$
$$591+87 \approx 600+90 = 690.$$
$$273+495 \approx 300+500 = 800.$$

Then, we compare these estimates to the values on the right. Since 500 is closest to 478, we pair $186+292$ and 478. Since 690 is closest to 678, we pair $591+87$ and 678. Since 800 is closest to 768, we pair $273+495$ and 768.

We connect these pairs without crossing wires as shown:

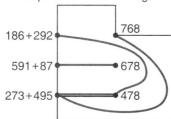

63. We estimate the products on the left:
$$442 \times 13 \approx 400 \times 10 = 4,000.$$
$$64 \times 14 \approx 60 \times 10 = 600.$$
$$873 \times 12 \approx 900 \times 10 = 9,000.$$

Then, we compare these estimates to the values on the right. Since 4,000 is closest to 5,746, we pair 442×13 and 5,746. Since 600 is closest to 896, we pair 64×14 and 896. Since 9,000 is closest to 10,476, we pair 873×12 and 10,476.

We connect these pairs without crossing wires as shown:

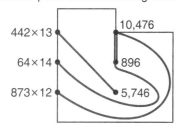

64. We first estimate the sums on the left:
$$3\tfrac{3}{5}+1\tfrac{1}{2} \approx 4+2 = 6.$$
$$5\tfrac{2}{3}+7\tfrac{5}{6} \approx 6+8 = 14.$$
$$6\tfrac{3}{4}+2\tfrac{1}{8} \approx 7+2 = 9.$$

Then, we compare these estimates to the values on the right. Since 6 is closest to $5\tfrac{1}{10}$, we pair $3\tfrac{3}{5}+1\tfrac{1}{2}$ and $5\tfrac{1}{10}$. Since 14 is closest to $13\tfrac{1}{2}$, we pair $5\tfrac{2}{3}+7\tfrac{5}{6}$ and $13\tfrac{1}{2}$. Since 9 is closest to $8\tfrac{7}{8}$, we pair $6\tfrac{3}{4}+2\tfrac{1}{8}$ and $8\tfrac{7}{8}$.

We connect these pairs without crossing wires as shown:

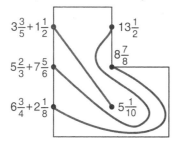

65. We estimate the products on the left:
$$3\tfrac{3}{5} \times 1\tfrac{2}{3} \approx 4 \times 2 = 8.$$
$$1\tfrac{7}{9} \times 5\tfrac{5}{8} \approx 2 \times 6 = 12.$$
$$8\tfrac{3}{4} \times 2\tfrac{2}{5} \approx 9 \times 2 = 18.$$

Then, we compare these estimates to the values on the right. Our estimate of 8 is close to both 6 and 10, but neither of the other estimates is close to 6. So, we pair $3\tfrac{3}{5} \times 1\tfrac{2}{3}$ and 6. Since 12 is closest to 10, we pair $1\tfrac{7}{9} \times 5\tfrac{5}{8}$ and 10. Since 18 is closest to 21, we pair $8\tfrac{3}{4} \times 2\tfrac{2}{5}$ and 21.

We connect these pairs without crossing wires as shown:

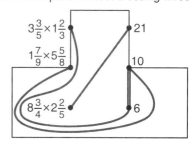

66. We estimate the products on the left:
$$168 \times 38 \approx 170 \times 40 = 6,800.$$
$$97 \times 202 \approx 100 \times 200 = 20,000.$$
$$904 \times 11 \approx 900 \times 10 = 9,000.$$

Then, we compare these estimates to the values on the right. Since 6,800 is closest to 6,384, we pair 168×38 and 6,384. Since 20,000 is closest to 19,594, we pair 97×202 and 19,594. Since 9,000 is closest to 9,944, we pair 904×11 and 9,944.

We connect these pairs without crossing wires as shown:

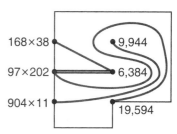

67. We estimate the products on the left:
$$18 \times 136 \approx 20 \times 140 = 2,800.$$
$$32 \times 54 \approx 30 \times 50 = 1,500.$$
$$97 \times 34 \approx 100 \times 34 = 3,400.$$

Then, we compare these estimates to the values on the right. Our estimate 2,800 is close to both 2,448 and 3,298. However, since neither of the other estimates is close to 2,448, we pair 18×136 and 2,448. Since 1,500 is closest to 1,728, we pair 32×54 and 1,728. Since 3,400 is closest to 3,298, we pair 97×34 and 3,298.

We connect these pairs without crossing wires as shown:

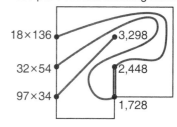

68. We estimate the products on the left:

$13 \times 87 \approx 10 \times 90 = 900$.
$49 \times 49 \approx 50 \times 50 = 2,500$.
$61 \times 71 \approx 60 \times 70 = 4,200$.

Then, we compare these estimates to the values on the right. Since 900 is closest to 1,131, we pair 13×87 and 1,131. Since 2,500 is closest to 2,401, we pair 49×49 and 2,401. Since 4,200 is closest to 4,331, we pair 61×71 and 4,331.

We connect these pairs without crossing wires as shown:

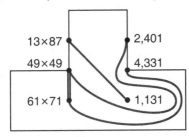

69. We estimate the products on the left:

$77 \times 37 \approx 80 \times 40 = 3,200$.
$63 \times 63 \approx 60 \times 60 = 3,600$.
$91 \times 19 \approx 90 \times 20 = 1,800$.

Then, we compare these estimates to the values on the right. Since 3,200 is closest to 2,849, we pair 77×37 and 2,849. Since 3,600 is closest to 3,969, we pair 63×63 and 3,969. Since 1,800 is closest to 1,729, we pair 91×19 and 1,729.

We connect these pairs without crossing wires as shown:

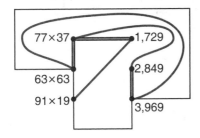

70. We rewrite each fraction on the left as a mixed number and estimate the sums:

$\frac{7}{4} + \frac{19}{8} = 1\frac{3}{4} + 2\frac{3}{8} \approx 2 + 2 = 4$.
$\frac{11}{4} + \frac{57}{8} = 2\frac{3}{4} + 7\frac{1}{8} \approx 3 + 7 = 10$.
$\frac{41}{4} + \frac{37}{8} = 10\frac{1}{4} + 4\frac{5}{8} \approx 10 + 5 = 15$.

Then, we rewrite each fraction on the right as a mixed number and estimate the sums:

$\frac{31}{4} + \frac{57}{8} = 7\frac{3}{4} + 7\frac{1}{8} \approx 8 + 7 = 15$.
$\frac{23}{8} + \frac{5}{4} = 2\frac{7}{8} + 1\frac{1}{4} \approx 3 + 1 = 4$.
$\frac{9}{2} + \frac{43}{8} = 4\frac{1}{2} + 5\frac{3}{8} \approx 5 + 5 = 10$.

Then, we compare the estimates from the left side to the estimates from the right side. Each estimate of a sum on the left has a matching estimate for a sum on the right.

So, we pair sums that have the same estimates.

$\frac{7}{4} + \frac{19}{8}$ and $\frac{23}{8} + \frac{5}{4}$,
$\frac{11}{4} + \frac{57}{8}$ and $\frac{9}{2} + \frac{43}{8}$,
$\frac{41}{4} + \frac{37}{8}$ and $\frac{31}{4} + \frac{57}{8}$.

We connect these pairs with wires that do not cross:

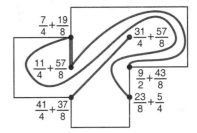

In fact, $\frac{7}{4} + \frac{19}{8}$ and $\frac{23}{8} + \frac{5}{4}$ are both equal to $\frac{33}{8}$ (or $4\frac{1}{8}$), $\frac{11}{4} + \frac{57}{8}$ and $\frac{9}{2} + \frac{43}{8}$ are both equal to $\frac{79}{8}$ (or $9\frac{7}{8}$), and $\frac{41}{4} + \frac{37}{8}$ and $\frac{31}{4} + \frac{57}{8}$ are both equal to $\frac{119}{8}$ (or $14\frac{7}{8}$).

71. First, we estimate the products on the left:

$88 \times 179 \approx 90 \times 200 = 18,000$.
$78 \times 77 \approx 80 \times 80 = 6,400$.
$247 \times 9 \approx 250 \times 10 = 2,500$.

Then, we estimate the products on the right:

$39 \times 57 \approx 40 \times 60 = 2,400$.
$22 \times 716 \approx 20 \times 700 = 14,000$.
$462 \times 13 \approx 500 \times 10 = 5,000$.

Then, we compare the estimates from the left side to the estimates from the right side. Since 18,000 is closest to 14,000, we pair 88×179 and 22×716. Since 6,400 is closest to 5,000, we pair 78×77 and 462×13. Since 2,500 is closest to 2,400, we pair 247×9 and 39×57.

We connect these pairs without crossing wires as shown:

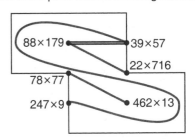

In fact, $88 \times 179 = 22 \times 716 = 15,752$, $78 \times 77 = 462 \times 13 = 6,006$, and $247 \times 9 = 39 \times 57 = 2,223$.

ESTIMATION
Over and Underestimating   58-59

72. Since $11\frac{17}{20}$ is less than 12 and $14\frac{4}{7}$ is less than 15, $11\frac{17}{20} + 14\frac{4}{7}$ is **less than** $12 + 15 = 27$.

In fact, $11\frac{17}{20} + 14\frac{4}{7} = 26\frac{59}{140}$, which is less than 27.

73. Since 23 is greater than 20, and 31 is greater than 30, 23×31 is **greater than** $20 \times 30 = 600$.

In fact, $23 \times 31 = 713$, which is greater than 600.

74. Since 13 is less than $13\frac{2}{9}$, and 2 is less than $2\frac{3}{8}$, $13\times2=26$ is less than $13\frac{2}{9}\times2\frac{3}{8}$. Therefore, 26 is an **underestimate** of the product $13\frac{2}{9}\times2\frac{3}{8}$.

In fact, $13\frac{2}{9}\times2\frac{3}{8}=31\frac{29}{72}$.

75. Since 6,000 is more than 5,725, and 19,000 is more than 18,842, we know that $6,000+19,000=25,000$ is more than $5,725+18,842$. So, 25,000 is an **overestimate** of $5,725+18,842$.

In fact, $5,725+18,842=24,567$.

76. When estimating a product (or sum) of two numbers, if both numbers are rounded up, the resulting estimate is always an overestimate. If both numbers are rounded down, the resulting estimate is always an underestimate.

However, **if one number is rounded up, and the other number is rounded down, the result can be an overestimate or an underestimate.** For example, if Norton estimates $7\times104\approx10\times100=1,000$, then his estimate is an overestimate, since $7\times104=728$. However, if Norton estimates $97\times14\approx100\times10=1,000$, then his estimate is an underestimate, since $97\times14=1,358$.

77. Since 85 is less than 90 and 17 is less than 20, we have 85×17 $\bigodot{<}$ 90×20.

In fact, $85\times17=1,445$, which is less than $90\times20=1,800$.

78. Since 8,519 is less than 9,000 and 4,672 is less than 5,000, we have $8,519+4,672$ $\bigodot{<}$ $9,000+5,000$.

In fact, $8,519+4,672=13,191$, which is less than $9,000+5,000=14,000$.

79. Since 461 is greater than 460 and 11 is greater than 10, we have 461×11 $\bigodot{>}$ 460×10.

In fact, $461\times11=5,071$, which is greater than $460\times10=4,600$.

80. We round 97 to 100 and 9 to 10, then estimate that $97\times9\approx100\times10=1,000$. Since we estimated the product by rounding both numbers up, our resulting estimate is an overestimate. So, we have 97×9 $\bigodot{<}$ $1,000$.

In fact, $97\times9=873$, which is less than 1,000.

81. We round $51\frac{5}{13}$ to 51 and $16\frac{1}{8}$ to 16, then estimate that $51\frac{5}{13}+16\frac{1}{8}\approx51+16=67$. Since we estimated the sum by rounding both numbers down, our resulting estimate is an underestimate. So, we have $51\frac{5}{13}+16\frac{1}{8}$ $\bigodot{>}$ 67.

In fact, $51\frac{5}{13}+16\frac{1}{8}=67\frac{53}{104}$, which is greater than 67.

82. We round $7\frac{3}{4}$ to 8 and $6\frac{7}{9}$ to 7, then estimate that $7\frac{3}{4}\times6\frac{7}{9}\approx8\times7=56$. Since we estimated the product by rounding both numbers up, our resulting estimate is an overestimate. So, we have $7\frac{3}{4}\times6\frac{7}{9}$ $\bigodot{<}$ 56.

In fact, $7\frac{3}{4}\times6\frac{7}{9}=52\frac{19}{36}$, which is less than 56.

83. Since there are 12 inches in one foot and 5,280 feet in one mile, there are $12\times5,280$ inches in one mile. We round 12 down to 10 and 5,280 down to 5,000 to estimate that there are about $10\times5,000=50,000$ inches in one mile. Since both numbers were rounded down and then multiplied, our estimate is smaller than the actual answer. Therefore, $12\times5,280$ inches is greater than 50,000 inches, and **1 mile** is greater than 50,000 inches.

In fact, $12\times5,280=63,360$.

84. Lunch Lady Lydia baked 36×21 cookies. If we round each number to the nearest ten, we estimate that Lydia made about $40\times20=800$ cookies. However, since we rounded 36 up to 40 and 21 down to 20, it is not clear whether we have made an overestimate or an underestimate. So, we need a different approach.

Instead, we estimate 36×21 by multiplying $35\times20=700$. Since 35 is less than 36 and 20 is less than 21, 35×20 is less than 36×21 and our estimate is an underestimate. So, the number of cookies baked is more than 700. **Yes,** Lydia will have enough cookies to give each of 700 students at least one cookie.

In fact, $36\times21=756$.

85. The rectangular patio will be 9 blocks wide and 13 blocks long. So, 9×13 blocks will be needed to make the patio. Since each block weighs 23 pounds, 9×13 blocks will weigh $9\times13\times23$ pounds. If we round each number to the nearest ten to estimate the total weight, we get $9\times13\times23\approx10\times10\times20=2,000$ pounds. However, we cannot tell whether 2,000 pounds is an overestimate or an underestimate. So, we need a different approach.

To estimate $9\times13\times23$, we can first consider the number of blocks (9×13). We can quickly see that 9×13 is more than 100, since $9\times11=99$. So, Rosencrantz and Guildenstern will need more than 100 blocks. Each block weighs 23 pounds, so they will need more than $100\times23=2,300$ pounds of blocks. Since one truck can only carry 2,000 pounds of blocks, **one truck cannot deliver enough blocks to build the patio.**

In fact, $9\times13\times23=2,691$.

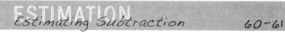

86. 501 is greater than 500, and 299 is less than 300. So, 501 and 299 are farther apart than 500 and 300. Therefore, $501-299$ is **greater** than $500-300=200$.

In fact, $501-299=202$.

87. $12\frac{3}{4}$ is less than 13, and $6\frac{1}{5}$ is greater than 6. So, $12\frac{3}{4}$ and $6\frac{1}{5}$ are closer together than 13 and 6. Therefore, $12\frac{3}{4}-6\frac{1}{5}$ is **less** than $13-6=7$.

In fact, $12\frac{3}{4}-6\frac{1}{5}=6\frac{11}{20}$.

88. 1,629 is greater than 1,600, and 976 is less than 1,000. So, 1,629 and 976 are farther apart than 1,600 and 1,000. Therefore, 1,629−976 is **greater** than 1,600−1,000 = 600.

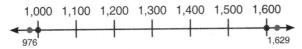

In fact, 1,629−976 = 653.

89. When estimating a difference, if the larger number is rounded up and the smaller number is rounded down, then the rounded numbers are farther apart than the original numbers. Therefore, the estimate of their difference is an overestimate.

If the larger number is rounded down and the smaller number is rounded up, then the rounded numbers are closer together than the original numbers. Therefore, the estimate of their difference is an underestimate.

However, if both numbers are rounded up or both are rounded down, then the result can be greater than or less than the actual difference.

Tara rounded $9\frac{1}{3}$ down to 9 and $2\frac{3}{7}$ down to 2 to estimate $9\frac{1}{3}-2\frac{3}{7} \approx 9-2 = 7$. **So, we cannot tell if 7 is an overestimate or underestimate without making an additional computation.**

90. 585 is less than 600, and 217 is more than 200. So, 585 and 217 are closer together than 600 and 200.

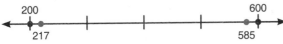

This means

$$585-217 \; \boxed{<} \; 600-200$$

In fact, 585−217 = 368, which is less than 600−200 = 400.

91. 1,200 is less than 1,219, and 600 is greater than 572. So, 1,219 and 572 are farther apart than 1,200 and 600.

This means

$$1,219-572 \; \boxed{>} \; 1,200-600$$

In fact, 1,219−572 = 647, which is greater than 1,200−600 = 600.

92. 1,491 is less than 1,500, and 718 is greater than 700. So, 1,491 and 718 are closer together than 1,500 and 700.

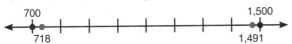

This means

$$1,491-718 \; \boxed{<} \; 1,500-700$$

In fact, 1,491−718 = 773, which is less than 1,500−700 = 800.

93. We round 907 to 900 and 587 to 600, then estimate that 907−587 ≈ 900−600 = 300. Since 900 is less than 907 and 600 is greater than 587, we know 900 and 600 are closer together than 907 and 587.

This means that our estimate is an underestimate. So,

$$300 \; \boxed{<} \; 907-587$$

In fact, 907−587 = 320, which is greater than 300.

94. We round $41\frac{7}{13}$ to 42 and $28\frac{1}{8}$ to 28, then estimate that $41\frac{7}{13}-28\frac{1}{8} \approx 42-28 = 14$. Since $41\frac{7}{13}$ is less than 42 and $28\frac{1}{8}$ is greater than 28, we know $41\frac{7}{13}$ and $28\frac{1}{8}$ are closer together than 42 and 28.

This means that our estimate is an overestimate. So,

$$14 \; \boxed{>} \; 41\frac{7}{13}-28\frac{1}{8}$$

In fact, $41\frac{7}{13}-28\frac{1}{8} = 13\frac{43}{104}$, which is less than 14.

95. We round $15\frac{2}{3}$ to 16 and $6\frac{2}{7}$ to 6, then estimate that $15\frac{2}{3}-6\frac{2}{7} \approx 16-6 = 10$. Since $15\frac{2}{3}$ is less than 16 and $6\frac{2}{7}$ is greater than 6, we know $15\frac{2}{3}$ and $6\frac{2}{7}$ are closer together than 16 and 6.

This means that our estimate is an overestimate. So,

$$15\frac{2}{3}-6\frac{2}{7} \; \boxed{<} \; 10$$

In fact, $15\frac{2}{3}-6\frac{2}{7} = 9\frac{8}{21}$, which is less than 10.

96. Adding or subtracting the same amount from two numbers does not change their difference. This is easiest to see on the number line:

When Yerg adds 11 to both 813 and 189, he does not change the difference. His new numbers are the same distance apart as his original numbers.
So, 813−189 = 824−200 = 624. **Yerg's answer is exact, not an estimate.** (Yerg's method is also an excellent strategy for subtraction.)

97. Instead of rounding to the nearest 10 or 100, we estimate 947 and 658 with numbers whose difference is easy to compute: 947 is very close to 950, and 658 is very close to 650. So, to make a close estimate of 947−658, we compute 950−650 = 300. Our estimate tells us 947−658 is closer to **300** than to 200.

In fact, 947−658 = 289, which is closer to 300 than to 200.

98. Timmy's estimate of the larger number is greater than the actual number. Timmy's estimate of the smaller number is less than the actual number. So, the numbers that Timmy subtracts are farther apart than the original numbers. Timmy's estimate of the difference is **more than** the actual difference.

99. Timmy's estimate of the larger number is 7 more than the actual number, and his estimate of the smaller number is 41 less than the actual number. The rounded numbers are farther apart than the actual numbers by $41+7=48$.

So, when Timmy subtracts the rounded numbers and gets 700, the difference is 48 more than the actual difference:

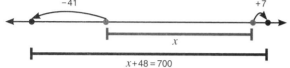

The actual difference between Timmy's numbers is $700-48=\textbf{652}$.

ESTIMATION
Estimating Division 62-63

For each problem below, you may have made different estimates than those given.

100. Since $90\times9=810$, we know that $810\div9=90$.
816 is close to 810, so we estimate $816\div9\approx810\div9=\textbf{90}$.

— *or* —

Since $80\times10=800$, we know that $800\div10=80$.
816 is close to 800, and 9 is close to 10, so we estimate $816\div9\approx800\div10=\textbf{80}$.

The actual value of $816\div9$ is $90\frac{2}{3}$. So, any estimate **from 80 to 100** is reasonable.

101. Since $40\times8=320$, we know that $320\div8=40$.
333 is close to 320, so we estimate $333\div8\approx320\div8=\textbf{40}$.

The actual value of $333\div8$ is $41\frac{5}{8}$. So, any estimate **from 30 to 50** is reasonable.

102. Since $34\frac{1}{9}$ is close to 34 and $6\frac{7}{8}$ is close to 7,

$34\frac{1}{9}\div6\frac{7}{8}$ is close to $34\div7$.

Since $5\times7=35$, we know that $35\div7=5$.

34 is close to 35, so we estimate $34\frac{1}{9}\div6\frac{7}{8}\approx35\div7=\textbf{5}$.

The actual value of $34\frac{1}{9}\div6\frac{7}{8}$ is $4\frac{476}{495}$.
So, any estimate **from 4 to 6** is reasonable.

103. We estimate $56,808\div789\approx56,000\div800=70$.
So, we expect that the quotient $56,875\div789$ is close to 70, which is a **2-digit** number.

In fact, $56,808\div789=72$, so the quotient is 2 digits long.

104. Since 6,237 is close to 6,000 and 11 is close to 10, we estimate $6,237\div11\approx6,000\div10=600$. Of the four answer choices, **567** is the only reasonable answer.

$$234 \quad \boxed{567} \quad 8,917 \quad 1,117$$

105. Since 12,321 is close to 12,000 and 37 is close to 40, we estimate $12,321\div37\approx12,000\div40=300$. Of the four answer choices, **333** is the only reasonable answer.

$$111 \quad \boxed{333} \quad 999 \quad 5,555$$

106. We can find the smallest possible quotient by dividing the smallest 6-digit number by 20. The smallest 6-digit number is 100,000. Since $20\times5,000=100,000$, we know that $100,000\div20=5,000$, which is a 4-digit number. So, **the quotient of a 6-digit number and 20 has at least 4 digits**.

107. Yerg's estimate is $800\div8=100$.
Plunk's estimate is $770\div10=77$.
Drew's estimate is $770\div7=110$.
In order from least to greatest, we have
77, 100, and 110.

108. We estimate each quotient, then match the two closest estimates:
$647\div82\approx640\div80=8$.
$6,478\div82\approx6,400\div80=80$.
$6,456\div8\approx6,460\div10=646$ *or* $6,456\div8\approx6,400\div8=800$.
$64,859\div821\approx64,000\div800=80$.

Our estimates of $6,478\div82$ and $64,859\div821$ are the same. The estimates of the other two quotients are not even close. So, $6,478\div82$ and $64,859\div821$ are the two with the same quotient.

$$647\div82 \quad \boxed{6,478\div82} \quad 6,456\div8 \quad \boxed{64,859\div821}$$

In fact, $6,478\div82=64,859\div821=79$.

109. To make the quotient as small as possible, we divide the smallest 4-digit number (1,000) by the largest 2-digit number (99). So, we wish to find the number of digits in $1,000\div99$.

We know that $1,000\div100=10$, which is a 2-digit number. If 1,000 is divided by a number *smaller* than 100, the result is *greater* than $1,000\div100=10$.

So, the quotient of $1,000\div99$ is at least 10 and therefore has at least 2 digits. Since the smallest possible quotient has at least 2 digits, the quotient of any 4-digit number and 2-digit number will have at least 2 digits. Therefore, **it is not possible to divide a 4-digit number by a 2-digit number and get a 1-digit quotient.**

ESTIMATION
Between 64-65

110. We estimate by adding the thousands first: $11,000+3,000=14,000$. Then, we estimate the remaining sum. Since $956+582$ is greater than 1,000 but less than 2,000, we know that $11,956+3,582$ is more than $14,000+1,000$, but less than $14,000+2,000$. So, $11,956+3,582$ is **between 15,000 and 16,000**.

— *or* —

11,956+3,582 is more than 11,500+3,500 = 15,000 because the numbers we added are smaller than the actual numbers. However, 11,956+3,582 is less than 12,000+4,000 = 16,000 because the numbers we added are larger than the actual numbers. So, 11,956+3,582 is **between 15,000 and 16,000**.

In fact, 11,956+3,582 = 15,538.

111. We estimate by adding the whole-number parts first: 4+7 = 11. Then, we estimate the sum of the fractional parts. Since $\frac{1}{6}$ and $\frac{3}{8}$ are both less than $\frac{1}{2}$, we know that $\frac{1}{6}+\frac{3}{8}$ is between 0 and 1. Therefore, $4\frac{1}{6}+7\frac{3}{8}$ is **between 11 and 12**.

In fact, $4\frac{1}{6}+7\frac{3}{8} = 11\frac{13}{24}$.

112. We know that 810÷9 = 90, and 900÷9 = 100. Since 857 is between 810 and 900, we know that 857÷9 is **between 90 and 100**.

In fact, $857÷9 = 95\frac{2}{9}$.

113. Since 439×57 is more than 400×50 = 20,000, but less than 500×60 = 30,000, we know that 439×57 is **between 20,000 and 30,000**.

In fact, 439×57 = 25,023.

For each of the following matching problems, you may have made different estimates to connect the same points.

114. $4\frac{7}{9}-2\frac{1}{5} \approx 5-2 = 3$, but since 5−2 gives us an overestimate, we connect $4\frac{7}{9}-2\frac{1}{5}$ to the only point that is less than 3.

$4\frac{7}{10}+2\frac{3}{5}$ is more than 4+2 = 6, so we connect $4\frac{7}{10}+2\frac{3}{5}$ to the only point that is more than 6.

Then, $1\frac{5}{6}×2\frac{7}{8}$ is less than 2×3 = 6. So, we connect $1\frac{5}{6}×2\frac{7}{8}$ to the point between 5 and 6 as shown below:

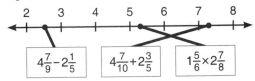

115. 596÷8 is more than 560÷8 = 70, but less than 640÷8 = 80. So, we connect 596÷8 to the point between 70 and 80.

273−175 is almost exactly 275−175 = 100, so we connect 273−175 to the point nearest 100.

372÷7 is more than 350÷7 = 50, but less than 420÷7 = 60. So, we connect 372÷7 to the point that is between 50 and 60 as shown below:

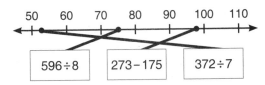

116. 1,356÷6 is more than 1,200÷6 = 200, but less than 1,800÷6 = 300. So, we connect 1356÷6 to the only point that is less than 300.

18×19 is less than 20×20 = 400, so we connect 18×19 to the remaining point that is less than 400.

216+226 is more than 200+200 = 400, so we connect 216+226 to the only point that is greater than 400 as shown below:

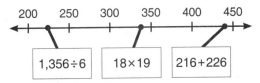

117. 25×23 is greater than 20×20 = 400, and less than 25×25 = (20×30)+25 = 625, so we connect 25×23 to the only point that is between 400 and 625.

38×9 is less than 40×10 = 400, so we connect it to the point that is less than 400.

254+481 is more than 250+450 = 700, so we connect 254+481 to the only point that is greater than 700 as shown below:

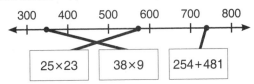

In the problems below, we estimate the value of each expression. If two estimates are close or the same, we look at whether each is an overestimate or underestimate.

118. 168+326 ≈ 170+330 = 500
547−198 ≈ 550−200 = 350
19×12 ≈ 20×10 = 200

So, we order the expressions from least to greatest: **19×12, 547−198, and 168+326**.

In fact, 19×12 = 228, and 547−198 = 349, and 168+326 = 494.

119. 3,116−2,338 ≈ 3,000−2,000 = 1,000
45×9 ≈ 50×10 = 500
316+246 ≈ 300+200 = 500

We get the same estimate for the second and third expressions:

45×9 is less than 50×10 = 500.
316+246 is more than 300+200 = 500.

We can order the expressions from least to greatest: **45×9, 316+246, and 3,116−2,338**.

In fact, 45×9 = 405, and 316+246 = 562, and 3,116−2,338 = 778.

120. $414+318 \approx 400+300 = 700$
$8,208 \div 36 \approx 8,000 \div 40 = 200$
$35 \times 19 \approx 35 \times 20 = 700$

We get the same estimate for the first and third expressions:

$414+318$ is more than $400+300 = 700$.
35×19 is less than $35 \times 20 = 700$.

We can order the expressions from least to greatest:
$8,208 \div 36$, 35×19, and $414+318$.

In fact, $8,208 \div 36 = 228$, and $35 \times 19 = 665$,
and $414+318 = 732$.

You may have made different estimates that allowed you to order the expressions correctly.

121. $456+465 \approx 450+450 = 900$.
$645+654 \approx 650+650 = 1,300$.
$546+564 \approx 550+550 = 1,100$.
So, we must visit the three expressions in this order:

Start ▶ $\boxed{\begin{array}{c}456\\+465\end{array}}$ ▶ $\boxed{\begin{array}{c}546\\+564\end{array}}$ ▶ $\boxed{\begin{array}{c}645\\+654\end{array}}$ Finish

We can only accomplish this with the path shown below:

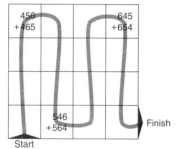

122. $31 \times 27 \approx 30 \times 30 = 900$.
$13 \times 39 \approx 10 \times 40 = 400$.
$11 \times 23 \approx 10 \times 20 = 200$.

So, we must visit the three expressions in this order:

Start ▶ $\boxed{\begin{array}{c}11\\ \times 23\end{array}}$ ▶ $\boxed{\begin{array}{c}13\\ \times 39\end{array}}$ ▶ $\boxed{\begin{array}{c}31\\ \times 27\end{array}}$ Finish

We can only accomplish this with the path shown below:

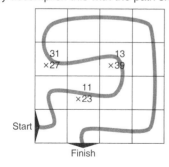

123. $258-147 \approx 250-150 = 100$.
$915-575 \approx 900-600 = 300$.
$659-112 \approx 650-100 = 550$.

So, we must visit the three expressions in this order:

Start ▶ $\boxed{\begin{array}{c}258\\-147\end{array}}$ ▶ $\boxed{\begin{array}{c}915\\-575\end{array}}$ ▶ $\boxed{\begin{array}{c}659\\-112\end{array}}$ Finish

We can only accomplish this with the path shown below:

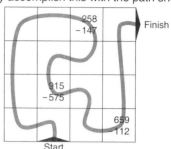

124. $145 \div 7 \approx 140 \div 7 = 20$.
$64 \div 5 \approx 60 \div 5 = 12$.
$59 \div 9 \approx 54 \div 9 = 6$.
$158 \div 10 \approx 160 \div 10 = 16$.
So, we must visit the four expressions in this order:

Start ▶ $\boxed{\begin{array}{c}59\\ \div 9\end{array}}$ ▶ $\boxed{\begin{array}{c}64\\ \div 5\end{array}}$ ▶ $\boxed{\begin{array}{c}158\\ \div 10\end{array}}$ ▶ $\boxed{\begin{array}{c}145\\ \div 7\end{array}}$ Finish

We can only accomplish this with the path shown below:

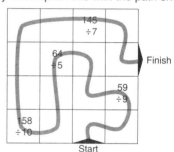

125. $46 \times 78 \approx 50 \times 80 = 4,000$.
$22 \times 12 \approx 20 \times 10 = 200$.
$82 \times 7 \approx 80 \times 7 = 560$.
$53 \times 18 \approx 50 \times 20 = 1,000$.

So, we must visit the four expressions in this order:

Start ▶ $\boxed{\begin{array}{c}22\\ \times 12\end{array}}$ ▶ $\boxed{\begin{array}{c}82\\ \times 7\end{array}}$ ▶ $\boxed{\begin{array}{c}53\\ \times 18\end{array}}$ ▶ $\boxed{\begin{array}{c}46\\ \times 78\end{array}}$ Finish

We can only accomplish this with the path shown below:

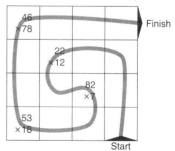

126. $996+632 \approx 1,000+600 = 1,600.$
$682+715 \approx 700+700 = 1,400.$
$844+319 \approx 850+300 = 1,150.$
$107+712 \approx 100+700 = 800.$

So, we must visit the four expressions in this order:

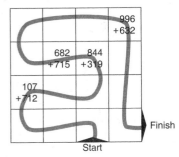

We can only accomplish this with the path shown below:

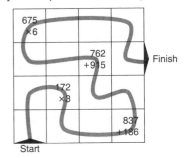

127. $675 \times 6 \approx 700 \times 6 = 4,200.$
$762+915 \approx 800+900 = 1,700.$
$172 \times 3 \approx 200 \times 3 = 600.$
$837+186 \approx 800+200 = 1,000.$

So, we must visit the four expressions in this order:

Start ▸ | 172 ×3 | 837 +186 | 762 +915 | 675 ×6 | ◂ Finish

We can only accomplish this with the path shown below:

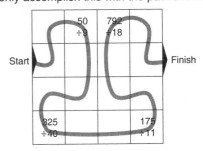

128. $50 \div 9 \approx 50 \div 10 = 5.$
$792 \div 18 \approx 800 \div 20 = 40.$
$325 \div 40 \approx 320 \div 40 = 8.$
$175 \div 11 \approx 170 \div 10 = 17.$

So, we must visit the four expressions in this order:

Start ▸ | 50 ÷9 | 325 ÷40 | 175 ÷11 | 792 ÷18 | ◂ Finish

We can only accomplish this with the path shown below:

129. $468+326 \approx 500+300 = 800.$
$226 \times 3 \approx 200 \times 3 = 600.$
$715 \div 6 \approx 700 \div 7 = 100.$
$33 \times 11 \approx 33 \times 10 = 330.$

So, we must visit the four expressions in this order:

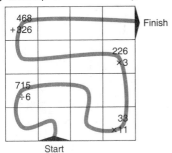

We can only accomplish this with the path shown below:

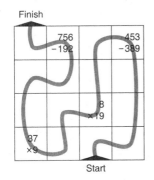

130. $37 \times 9 \approx 37 \times 10 = 370.$
$756-192 \approx 750-200 = 550.$
$453-389 \approx 450-400 = 50.$
$8 \times 19 \approx 8 \times 20 = 160.$

So, we must visit the four expressions in this order:

We can only accomplish this with the path shown below:

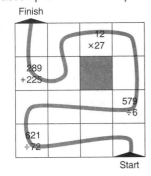

131. $12 \times 27 \approx 10 \times 30 = 300.$
$289+225 \approx 300+200 = 500.$
$579 \div 6 \approx 600 \div 6 = 100.$
$621 \div 72 \approx 630 \div 70 = 9.$

So, we must visit the four expressions in this order:

We can only accomplish this with the path shown below:

132. $921-389 \approx 900-400 = 500.$
$635+478 \approx 600+500 = 1,100.$
$758 \div 25 \approx 750 \div 25 = 30.$
$85 \times 758 \approx 90 \times 800 = 72,000.$

So, we must visit the four expressions in this order:

We can only accomplish this with the path shown below:

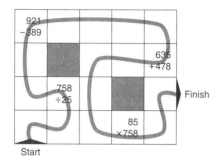

All the book prices are near a whole dollar amount except for The Call of the Basilisk ($3.50). We round the prices of the other books to the nearest whole dollar:

Where the Wild Things Are	$6.85 \approx $7	A Tale of Two Yetis	$1.95 \approx $2
Diary of a Wimpy Cyclops	$5.95 \approx $6	The Rancor in the Hat	$4.99 \approx $5
All the Pretty Jackalopes	$9.95 \approx $10	The Call of the Basilisk	$3.50

133. We estimate the least amount of money necessary to buy three books on this list. The three least expensive books on the list are A Tale of Two Yetis ($1.95), The Call of the Basilisk ($3.50), and The Rancor in the Hat ($4.99). The total cost of these books ($1.95+$4.99+$3.50) is less than 2+5+4 = 11 dollars. **So, Lizzie can buy three different books for $11: A Tale of Two Yetis, The Call of the Basilisk, and The Rancor in the Hat.**

The actual cost of these three books is $1.95+$3.50+$4.99 = $10.44. No other set of 3 books on this list can be bought with $11 or less.

134. All of the books' prices are near a whole dollar amount except for The Call of the Basilisk ($3.50), so we guess that Grogg bought Call of the Basilisk and another book.

We estimate that $9.45 is about 6 dollars more than $3.50. So, Grogg's other book costs about $6.

The only book that costs about $6 is Diary of a Wimpy Cyclops. So, we guess that Grogg bought **Diary of a Wimpy Cyclops ($5.95) and The Call of the Basilisk ($3.50)**.

No other pair of books on this list costs between $9 and $10. In fact, $5.95+$3.50 = $9.45.

135. All of the books are near a whole dollar amount except for The Call of the Basilisk ($3.50). So, we first estimate the sum of the costs of the five books near a whole dollar amount: 7+6+10+2+5 = 30 dollars. Our estimate is a slight overestimate because we rounded these prices up by a small number of cents. Then we add the cost of The Call of the Basilisk ($3.50) to our sum of $30.
So, we estimate that the total cost of the books is a little less than 30+3 = 33 dollars and 50 cents. So, the cost is between **$30 and $35**.

$20 and $25	$25 and $30	($30 and $35)	$35 and $40

In fact, the total cost of all 6 books is $33.19, which is between $30 and $35.

136. If five copies of a book cost $34.25, then one copy costs $34.25 \div 5 \approx 35 \div 5 = 7$ dollars. **Where the Wild Things Are ($6.85)** costs about $7.

To be safe, we check the cost of 5 copies of the other books that cost about $7:
Five copies of Diary of a Wimpy Cyclops cost 5×5.95 dollars, which is a little less than $5 \times 6 = 30$ dollars.
Five copies of All the Pretty Jackalopes cost 5×9.95 dollars, which is more than $5 \times 9 = 45$ dollars.
Five copies of any other book on the list cost less than $30 or more than $45.

137. To make an eyeball estimate of the number of ounces of paint required, we draw lines that split the shape up into pieces that are about the same size as the square:

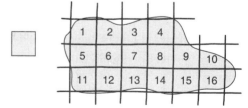

The shape on the right covers about the same area as 16 small squares. So, about **16 ounces** of paint will be needed to paint the shape.

Any estimate **from 12 to 20 ounces** is reasonable.

138. To estimate the number of dots, we split the set of dots into groups that are about the same size. We count the number of dots in the group, then multiply by the number of groups.

For example, we have split the dots below into 9 groups that are about the same size. We count 20 dots in the shaded group below.

So, we estimate that there are about $9 \times 20 = $ **180** dots.

In fact, there are 182 dots. Any estimate **from 130 to 250 dots** is reasonable.

139. The distance between East Sasquatch and Cyclopton along Highway 90 looks about three times the distance between Phoenix and East Sasquatch:

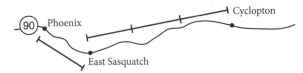

So, the distance from Phoenix to Cyclopton is about four times the distance from Phoenix to East Sasquatch. Since the distance from Phoenix to East Sasquatch is about 17 miles, the distance from Phoenix to Cyclopton is about $17 \times 4 = 68$ miles, which we round to about **70 miles**.

Any estimate from **50 to 80 miles** is reasonable.

140. As shown in the previous problem, the distance from East Sasquatch to Cyclopton is about three times the distance from Phoenix to East Sasquatch. Since it takes about 60 minutes to drive from East Sasquatch to Cyclopton, it will take about $60 \div 3 = $ **20 minutes** to drive from Phoenix to East Sasquatch.

Any estimate **from 15 to 30 minutes** is reasonable.

141. We use our knowledge of clocks to make an eyeball estimate of the time.

Since the hour hand is between the 6 and 7, we know that the time is between 6:00 and 7:00. Since the hour hand is closer to the 7 than to the 6, the time is between 6:30 and 7:00. So, we estimate that the time is about **6:45**.

The actual time displayed by the hour hand is 6:48. Any estimate **from 6:35 to 6:55** is reasonable.

142. We estimate that the lighthouse is about as tall as 7 Groggs, as shown below.

So, we estimate that the lighthouse is about $5 \times 7 = $ **35 feet** tall.

Any estimate **from 25 to 50 feet** is reasonable.

143. If all of the sides of the triangle were the same length, each side would be $12 \div 3 = 4$ centimeters long. Since we are estimating the side length of a scalene triangle, the length of the short side is shorter than 4 cm. So, we guess that the short side is about **3 cm** long.

Any estimate **from 2 to 4 cm** is reasonable.

144. We estimate that the line is about as long as five quarters. Since a quarter is about one inch across, we estimate that the line is about **5 inches** long.

In fact, the line is $4\frac{3}{4}$ inches long.

Any estimate **from 4 to 6 inches** is reasonable.

145. There are 8 stacks of nickels: two short, one medium, and five large. We count a short stack to find that it contains 10 coins. So each short stack is worth $5 \times 10 = 50$ cents. Together, the two short stacks are worth $50 + 50 = 100$ cents ($1). Then, the one medium stack looks about twice as tall as a short stack, so we estimate that the medium stack is worth as much as two short stacks: $1. Finally, each large stack looks about twice as tall as the medium stack, so we estimate that a large stack is worth as much as two medium stacks: $1 \times 2 = \$2$. Five large stacks are worth $\$2 \times 5 = \$10$. All together, we estimate that the stacks are worth $\$1 + \$1 + \$10 = $ **\$12**.

Any estimate **from \$10 to \$15** is reasonable.

146. One pound of beastberries costs about \$3. Reading the scale, we see that this bag weighs about 2 pounds. So, we estimate that this bag costs about $3 \times 2 = $ **6 dollars**.

Any estimate **from \$5 to \$8** is reasonable.

147. One pound of beastberries costs about \$3. Reading the scale, we see that this bag weighs just over $3\frac{1}{4}$ pounds. So, we estimate that this bag costs a little more than $3 \times 3 = 9$ dollars. Our eyeball estimate for the cost of this bag is **\$9 or \$10**.

Any estimate **from \$8 to \$12** is reasonable.

Our answers to Fermi problems are rough guesses, so we include a range of reasonable answers.

There are many different ways to arrive at a reasonable estimate for each problem below!

148. We estimate that an apple is about 3 inches long, 3 inches wide, and 3 inches tall.

A standard bathtub is about 60 inches (5 ft) long, 24 inches wide (2 ft), and 15 inches (1 ft 3 in) deep.

Since each apple is about 3 inches long, and a bathtub is 60 inches long, we can place about $60 \div 3 = 20$ apples along the length of the bathtub.

Since each apple is about 3 inches wide, and a bathtub is about 24 inches wide, we can place about $24 \div 3 = 8$ apples along the width of the bathtub.

So, we can cover the bottom of a bathtub with about $20 \times 8 = 160$ apples.

Then, since each apple is about 3 inches tall, and a bathtub is about 15 inches deep, we can stack about $15 \div 3 = 5$ apples to reach the depth of the tub. This gives us 5 layers of 160 apples for a total of about $5 \times 160 = 800$ apples. So, we estimate that it will take **about 800 apples** to fill a bathtub.

Any estimate **from 400 to 1,600 apples** is reasonable. A very large tub may hold even more than 1,600 very small apples.

149. Most people cut their fingernails every couple of weeks, but each nail only grows a small fraction of an inch in that time. It takes about a year for most people to grow a fingernail that is an inch long. Some fingernails grow slower, some grow faster. So, if you live to be 80 years old, your fingernails will grow about 80 inches all together. There are 12 inches in a foot, so 5 feet equals 60 inches, 6 feet equals 72 inches, and 7 feet equals 84 inches. So, 80 inches is a little less than 7 feet. If you could manage to grow your fingernails for your whole life, they would grow about **7 feet long**.

The Guinness World Record for fingernail length is over 10 feet! Any estimate **from 5 to 20 feet** is reasonable.

150. Count the number of breaths you take in one minute. Most people take a breath about every 4 to 6 seconds, or about 10-15 breaths per minute. There are 60 minutes in an hour and 24 hours in a day, so most people take between $10 \times 60 \times 24$ and $15 \times 60 \times 24$ breaths in a day.

$10 \times 60 \times 24 = 600 \times 24 \approx 600 \times 25 = 15,000.$

$15 \times 60 \times 24 = 900 \times 24 \approx 1,000 \times 24 = 24,000.$

So, we guess that most people take **between 15,000 and 24,000 breaths** each day.

An estimate **from 10,000 to 30,000 breaths** is reasonable.

151. An hour of television usually has about 15 minutes of commercials. Commercials last about 30 seconds each, which means that in one minute of commercials, there are about 2 commercials. So, there are about $2 \times 15 = 30$ commercials during every hour of television. There are 16 hours from 6 a.m. to 10 p.m., so there are about 16×30 commercials from 6 a.m. to 10 p.m.

Since $15 \times 30 = 450$ and $20 \times 30 = 600$, we guess that there are **between 450 and 600 commercials** on between 6 a.m. and 10 p.m. every day.

Any guess **between 200 and 1,000 commercials** is reasonable. Please do not watch 16 hours of television to find out.

152. A normal walking speed is about 2 miles per hour. This means that you could probably walk about two miles each hour. So, it would take about $2,800 \div 2 = 1,400$ hours to walk from New York to Los Angeles without ever stopping. There are 24 hours in one day, so we must divide $1,400 \div 24$ to figure out how many days of walking this is. We can estimate that 100 hours is about $100 \div 25 = 4$ days. There are 14 hundreds in 1,400, so 1,400 hours is about $14 \times 4 = 56$ days. Depending on how fast you walk, it would take **between 30 and 80 days** to walk without stopping from Los Angeles to New York.

Any estimate **from 20 to 100 days** is reasonable.

153. Please do not try to count the number of words in this book! Some pages in this book have lots of words, but others have very few. The hint and solution pages have many more words than the problem pages, so we will estimate hints and solution pages and problem pages separately.

On a hint or solution page like this one, there are about 10 words for every line of text. In each column, there are about 50 lines of text per column, and there are 2 columns. So, each hint and solution page has about $10 \times 50 \times 2 = 1,000$ words. There are about 50 solution pages for a total of 50,000 words in the solutions.

There are lots of different types of problem pages. The pages with lots of text balance out the pages with lots of numbers or diagrams that have very little text. We guess that a typical problem page has about 100 words. There are about 100 pages of problems, for a total of 10,000 words in the problems pages.

So, we guess that there are about $50,000 + 10,000 = $ **60,000 words** in this book.

A guess **from 10,000 to 100,000 words** is reasonable. (We don't actually know the real answer.)

AREA
Square Units
77-79

1. The area of the 4 cm by 4 cm square is
$4 \times 4 = $ **16 square centimeters (16 sq cm)**.

2. The area of the 3 cm by 5 cm rectangle is
$3 \times 5 = $ **15 square centimeters (15 sq cm)**.

3. The area of the 5 cm by 5 cm square is
$5 \times 5 = $ **25 square centimeters (25 sq cm)**.

4. The area of the 6 cm by 4 cm rectangle is
$6 \times 4 = $ **24 square centimeters (24 sq cm)**.

5. The area of the 8 ft by 8 ft square is
$8 \times 8 = $ **64 square feet (64 sq ft)**.

6. The area of the 10 m by 15 m rectangle is
$10 \times 15 = $ **150 square meters (150 sq m)**.

7. The area of the 20 yd by 27 yd rectangle is
$20 \times 27 = $ **540 square yards (540 sq yd)**.

8. The area of the 50 km by 41 km rectangle is
$50 \times 41 = 50 \times (40+1) = (50 \times 40) + (50 \times 1)$
$= 2,000 + 50 = $ **2,050 square kilometers (2,050 sq km)**.

9. The area of the 7 mi by 14 mi rectangle is
$7 \times 14 = $ **98 square miles (98 sq mi)**.

10. The area of the 60 inch by 99 inch rectangle is
$60 \times 99 = 60 \times (100-1) = (60 \times 100) - (60 \times 1)$
$= 6,000 - 60 = $ **5,940 square inches (5,940 sq in)**.

11. The area of a square with 9-cm sides is
$9 \times 9 = $ **81 square centimeters (81 sq cm)**.

12. The area of the rectangular floor is
$40 \times 30 = $ **1,200 square feet (1,200 sq ft)**.

13. The area of one rectangular tabletop is
$60 \times 30 = $ **1,800 square inches (1,800 sq in)**.

14. The area of the rectangular field is
$30 \times 100 = $ **3,000 square yards (3,000 sq yd)**.

15. To find the height of a rectangle, we can divide the rectangle's area by its width. The area of this rectangle is 42 square inches, and its width is 6 inches. Therefore, its height is $42 \div 6 = 7$ inches.
The perimeter of a 6 inch by 7 inch rectangle is
$(6+7) + (6+7) = 13 + 13 = $ **26 inches (26 in)**.

16. The height of each congruent rectangle is equal to the side length of the square. The combined width of the three congruent rectangles is also equal to the side length of the square. So, the height of each rectangle is three times the rectangle's width. We look for a rectangle with a perimeter of 16 inches whose height is three times the width:

2 in

1 in

3 in

6 in

The perimeter of the 2 inch by 6 inch rectangle is
$(2+6) + (2+6) = 8 + 8 = 16$ inches. So, the side length of the square created by the attached rectangles is 6 inches. The area of a square with side length 6 inches is
$6 \times 6 = $ **36 square inches (36 sq in)**.

— or —

If we use w to represent the width of each congruent rectangle, then we can express the height (the side length of the square) as $w+w+w$, as shown:

We write an expression for the perimeter of each rectangle: $w+(w+w+w)+w+(w+w+w) = 8 \times w$.
The perimeter of each rectangle is 16 inches, so
$$8 \times w = 16 \text{ inches.}$$
Since $8 \times \boxed{2} = 16$, we have $w = 2$ inches. So, the width of each rectangle is 2 inches, and the height is $2+2+2 = 6$ inches. The side length of the square is equal to the height of one rectangle: 6 inches.

The area of a square with side length 6 inches is
$6 \times 6 = $ **36 square inches (36 sq in)**.

AREA
Adding Areas
80-81

17. Since $3 \times 3 = 9$, a square with an area of 9 square miles has side length 3 miles. So, the shaded square and the square below it both have side length 3 miles:

The side length of the large square is $3+3 = 6$ miles.

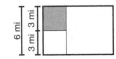

We calculate the area of each square, then add those areas to get the area of the rectangle they create:

$(3 \times 3) + (3 \times 3) + (6 \times 6) = 9 + 9 + 36 = $ **54 sq mi**.

— or —

After finding the side length of each square, we see that the large rectangle has height 6 miles and width $3+6 = 9$ miles. So, the area of the rectangle is $6 \times 9 = $ **54 sq mi**.

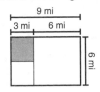

18. The combined width of 3 congruent squares is 6 inches. So, the side length of one square is $6 \div 3 = 2$ inches.

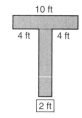

The area of a 2-inch square is $2 \times 2 = 4$ square inches. The shape is made from nine 2 inch by 2 inch squares. So, the total area of the shape is $9 \times 4 =$ **36 sq in**.

19. Attaching two rectangles along a side creates a rectilinear shape. The width of the shape is 10 feet. Since $4 + \boxed{2} + 4 = 10$, the unlabeled horizontal side must be 2 feet, as shown below:

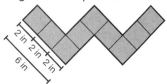

So, each rectangle is 2 ft by 10 ft, and the area of each rectangle is $2 \times 10 = 20$ square feet.

The "T" is made from 2 rectangles, each with an area of 20 square feet. The area of the "T" is $2 \times 20 =$ **40 sq ft**.

For each shape in the four problems that follow, we show one possible way to split the shape into rectangles. You may have split the shapes differently to calculate the same areas.

20. We split the shape into rectangles:

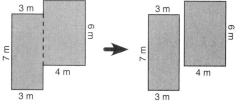

The area of the 7 m by 3 m rectangle is $7 \times 3 = 21$ sq m. The area of the 6 m by 4 m rectangle is $6 \times 4 = 24$ sq m. The total area of the rectilinear shape is $21 + 24 =$ **45 sq m**.

21. We split the shape into rectangles.

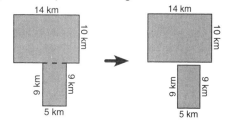

The area of the 10 km by 14 km rectangle is $10 \times 14 = 140$ sq km. The area of the 9 km by 5 km rectangle is $9 \times 5 = 45$ sq km. The total area of the rectilinear shape is $140 + 45 =$ **185 sq km**.

22. We split the shape into rectangles:

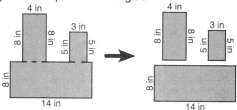

The area of the 8 inch by 4 inch rectangle is $8 \times 4 = 32$ sq in. The area of the 5 inch by 3 inch rectangle is $5 \times 3 = 15$ sq in. The area of the 8 inch by 14 inch rectangle is $8 \times 14 = 112$ sq in. The total area of the rectilinear shape is $32 + 15 + 112 =$ **159 sq in**.

23. We split the shape into rectangles:

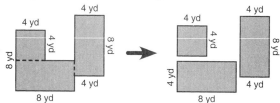

The area of the 4 yd by 4 yd square is $4 \times 4 = 16$ sq yd. The area of 4 yd by 8 yd rectangle is $4 \times 8 = 32$ sq yd. The area of the 8 yd by 4 yd rectangle is $8 \times 4 = 32$ sq yd. The total area of the rectilinear shape is $16 + 32 + 32 =$ **80 sq yd**.

24. The long side of each small rectangle has length 9 meters. Three short sides of the small rectangles equal the height of the large rectangle. So, the short side of a small rectangle has length $9 \div 3 = 3$ meters.

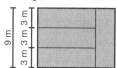

The area of each 3 meter by 9 meter rectangle is $3 \times 9 = 27$ sq m. The large rectangle is made of 4 congruent rectangles, each with area 27 sq m. So, the area of the large rectangle is $27 \times 4 =$ **108 sq m**.

— *or* —

After finding the side lengths of the small rectangles, we see that the width of the large rectangle is $9 + 3 = 12$ meters.

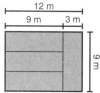

The area of a 9 meter by 12 meter rectangle is $9 \times 12 =$ **108 sq m**.

25. The area of the 8 inch by 10 inch rectangle is
$8 \times 10 = 80$ sq in. The area of the 4-inch square hole
is $4 \times 4 = 16$ sq in. The area of the remaining shape is
$80 - 16 =$ **64 sq in**.

26. We think of the figure as a large rectangle with a small
rectangle removed, as shown:

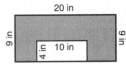

The area of the large rectangle is $20 \times 9 = 180$ sq in.
The area of the small rectangle is $4 \times 10 = 40$ sq in.
The area of the shape formed by removing the small
rectangle from the large one is $180 - 40 =$ **140 sq in**.

27. We think of the figure as a large square with a small
square removed, as shown:

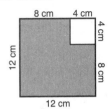

The area of the large square is $12 \times 12 = 144$ sq cm. The
area of the small square is $4 \times 4 = 16$ sq cm. The area of
the shape formed by removing the small square from the
large one is $144 - 16 =$ **128 sq cm**.

28. We think of the figure as a large rectangle with two
smaller rectangles removed, as shown:

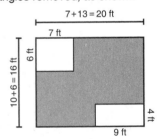

The area of the large rectangle is $16 \times 20 = 320$ sq ft.
The area of the 6 ft by 7 ft rectangle is $6 \times 7 = 42$ sq ft.
The area of the 4 ft by 9 ft rectangle is $4 \times 9 = 36$ sq ft.
The area of the shape formed by removing the two small
rectangles from the large one is $320 - 42 - 36 =$ **242 sq ft**.

29. The number of squares used to make the shape is three
less than the number of squares in a 4 square by 12
square rectangle. That is $(4 \times 12) - 3 = 48 - 3 = 45$ squares.

Each square has side length 2 cm, so the area of each
square is $2 \times 2 = 4$ sq cm.

The total area of 45 squares, each with an area of
4 sq cm, is $45 \times 4 =$ **180 sq cm**.

— *or* —

The side length of each square is 2 cm. So, the area of
this shape is the same as the area of a rectangle with
height $4 \times 2 = 8$ cm and width $12 \times 2 = 24$ cm, minus the
areas of three 2 cm by 2 cm squares.

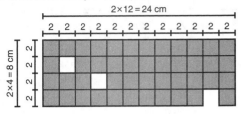

The area of an 8 cm by 24 cm rectangle is
$8 \times 24 = 192$ sq cm.
The area of each 2 cm by 2 cm square is $2 \times 2 = 4$ sq cm.
So, three 2 cm by 2 cm squares have a total area of
$3 \times 4 = 12$ sq cm.

The area of the shape formed by removing the three
squares from the large rectangle is $192 - 12 =$ **180 sq cm**.

30. The side length of the original square is 8 ft. So, the area
of the original square is $8 \times 8 = 64$ sq ft. The square is
split into four congruent triangles, so the area of each
triangle is $64 \div 4 = 16$ sq ft. When one triangle is removed,
the area of the remaining shape is $64 - 16 =$ **48 sq ft**.

31. Since the original shape was a square, the left, right, and
bottom sides are each 10 yards long. The total length
of the three horizontal sides on top is equal to the side
length of the square: 10 yards.

So, the total length of the bold sides in the diagram below
is $4 \times 10 = 40$ yards.

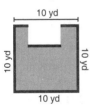

The perimeter of the entire shape is 46 yards.
So, the combined length of the two missing short vertical
sides is $46 - 40 = 6$ yards. The two sides are the opposite
sides of a rectangle, so they are the same length. Each
short vertical side is therefore $6 \div 2 = 3$ yards long.

The area of the original square is $10 \times 10 = 100$ square
yards. The area of the removed rectangle is $3 \times 5 = 15$
square yards. So, the area of the remaining shape is
$100 - 15 =$ **85 square yards**.

32. To find the area of the hole, we find the area of the large square created by attaching the quadrilaterals, then subtract the total area of the quadrilaterals. The area of the large square is $10 \times 10 = 100$ sq cm. Since the area of one quadrilateral is 17 sq cm, the area of the four congruent quadrilaterals is $17 \times 4 = 68$ sq cm. So, the area of the square hole is $100 - 68 = $ **32 sq cm.**

33. The combined width of the two smallest squares is the same as the width of the square above them. The bottom-left square has side length 3 inches, and the square above has side length 5 inches. Since $3 + \boxed{2} = 5$, the smallest square has side length 2 inches:

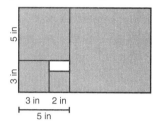

The height of the rectangle formed by attaching all the squares is $3 + 5 = 8$ inches, so the side length of the largest square is 8 inches.

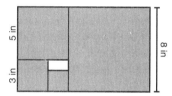

From here, we calculate the total shaded area by adding the area of each attached square.

The area of a 2 inch by 2 inch square is $2 \times 2 = 4$ sq in.
The area of a 3 inch by 3 inch square is $3 \times 3 = 9$ sq in.
The area of a 5 inch by 5 inch square is $5 \times 5 = 25$ sq in.
The area of a 8 inch by 8 inch square is $8 \times 8 = 64$ sq in.
The total area of the four squares is
$4 + 9 + 25 + 64 = $ **102 sq in.**

— *or* —

After finding the side length of each square, we calculate the total area of the rectangle created and then subtract the area of the small unshaded rectangle. The width of the rectangle formed by attaching all the squares is $8 + 5 = 13$ inches.

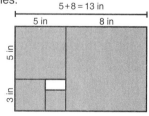

The area of an 8 inch by 13 inch rectangle is $8 \times 13 = 104$ sq in. The width of the unshaded rectangle the same as the square below it: 2 inches. The height of the unshaded rectangle is the difference between the height of the 3-inch square and the height of the 2-inch square: $3 - 2 = 1$ inch.

The area of a 1 inch by 2 inch rectangle is $1 \times 2 = 2$ sq in. So, the total shaded area is $104 - 2 = $ **102 sq in.**

34. To calculate the total shaded area, we begin by finding the area of the largest shaded square, with side length 7 cm: $7 \times 7 = 49$ sq cm.

Then, if we subtract the area of the unshaded square with side length 5 cm, we get the shaded area of the diagram below:

$49 - (5 \times 5) = 49 - 25 = 24$ sq cm.

Finally, we add the area of the shaded square with side length 3 cm to get the total shaded area of the diagram below:

$24 + (3 \times 3) = 24 + 9 = $ **33 sq cm.**

AREA
Tangrams   85-87

For clarity, we have shaded each tan as shown:

35. We arrange the 7 tans into the "Bunny" as shown:

36. We arrange the 7 tans into the "Zombie" as shown:

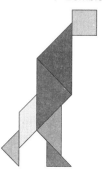

37. We arrange the 7 tans into the "Meerkat" as shown:

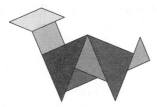

38. We arrange the 7 tans into the "Fish" as shown:

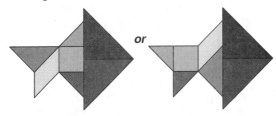

You may have found another solution.

39. We arrange the 7 tans into a triangle as shown:

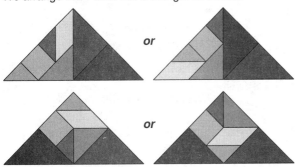

or

or

You may have found another solution.

40. We arrange the 7 tans into a rectangle as shown:

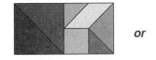

or

You may have found another solution.

41. We arrange the 7 tans into a square with a missing triangle as shown:

or

You may have found another solution.

42. We arrange the 7 tans into a quadrilateral with a missing quadrilateral as shown:

or

You may have found another solution.

43. The area of this right triangle is half the area of a rectangle with height 5 miles and width 12 miles:

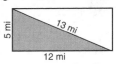

The area of a 5 mile by 12 mile rectangle is 5×12 = 60 sq mi. Half of 60 is 30, so the area of this triangle is **30 sq mi**.

44. The area of this right triangle is half the area of a rectangle with height 20 inches and width 21 inches:

The area of a 20 inch by 21 inch rectangle is 20×21 = 420 sq in. Half of 420 is 210, so the area of this triangle is **210 sq in**.

45. The area of this right triangle is half the area of a rectangle with height 8 yards and width 15 yards:

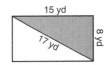

The area of a 8 yard by 15 yard rectangle is 8×15 = 120 sq yd. Half of 120 is 60, so the area of this triangle is **60 sq yd**.

46. The area of this right triangle is half the area of a rectangle with height 6 meters and width 8 meters:

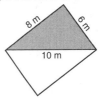

The area of a 6 meter by 8 meter rectangle is 6×8 = 48 sq m. Half of 48 is 24, so the area of this triangle is **24 sq m**.

47. This shape is made of two congruent right triangles. We can arrange the two congruent right triangles to make a rectangle, as shown.

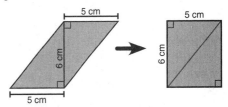

So, the total area of the shape created by attaching these two triangles is the same as the area of a 6 cm by 5 cm rectangle: 6×5 = **30 sq cm**.

48. This shape is made of two congruent right triangles. We can arrange the two congruent right triangles to make a rectangle, as shown.

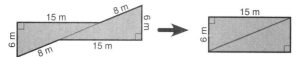

So, the total area of the shape created by attaching these two triangles is the same as the area of a 6 meter by 15 meter rectangle: $6 \times 15 =$ **90 sq m**.

49. This shape is made of four congruent triangles. We can arrange two congruent right triangles to make a rectangle, so we can arrange the *four* congruent triangles to make *two* rectangles, as shown:

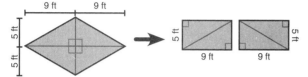

The total area of these four triangles is the same as the area of two 5 foot by 9 foot rectangles: $2 \times (5 \times 9) = 2 \times 45 =$ **90 sq ft**.

50. We can draw a line to split the shape into two right triangles as shown below. The area of this shape is the sum of the areas of the two triangles.

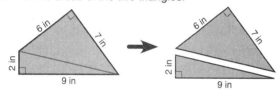

The area of the top triangle is half the area of a 6 inch by 7 inch rectangle. The area of a 6 inch by 7 inch rectangle is $6 \times 7 = 42$ sq in. Half of 42 is 21, so the area of the top triangle is 21 sq in.

The area of the bottom triangle is half the area of a 2 inch by 9 inch rectangle. The area of a 2 inch by 9 inch rectangle is $2 \times 9 = 18$ sq in. Half of 18 is 9, so the area of the bottom triangle is 9 sq in.

The total area of these two triangles is $21 + 9 =$ **30 sq in**.

51. We label all known horizontal lengths to find the length of the horizontal side of each triangle:

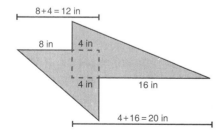

We label all known vertical lengths to find the length of the vertical side of each triangle:

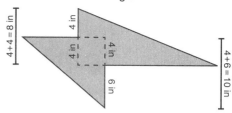

So, the area of the triangle in the upper-right is half the area of a rectangle with height 8 inches and width 20 inches. The area of an 8 inch by 20 inch rectangle is $8 \times 20 = 160$ sq in. Half of 160 is 80, so the area of the upper-right triangle is 80 sq in.

Subtracting the area of the 4-inch square gives the area of the shape below:

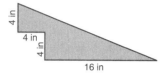

The area of this shape is $80 - (4 \times 4) = 80 - 16 = 64$ sq in.

The area of the lower-left triangle is half the area of a rectangle with height 10 inches and width 12 inches. The area of a 10 inch by 12 inch rectangle is $10 \times 12 = 120$ sq in. Half of 120 is 60, so the area of the lower-left triangle is 60 sq in.

Adding the area of the bottom-left triangle to the area of the upper-right triangle with the square removed gives us the total area of the shape:

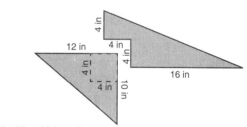

$64 + 60 =$ **124 sq in**.

You may have split each shape differently to add or subtract areas and arrive at the same total area.

52. We can split the shape into two triangles, as shown:

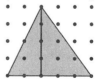

The triangle on the left is half of a 2 unit by 4 unit rectangle. The area of a 2 unit by 4 unit rectangle is 8 square units. Half of 8 is 4, so the area of the triangle on the left is 4 square units.

The triangle on the right is half of a 3 unit by 4 unit rectangle. The area of a 3 unit by 4 unit rectangle is $3\times4=12$ square units. Half of 12 is 6, so the area of the triangle on the right is 6 square units.

The total area of the shaded region is $4+6=$ **10 square units**.

— or —

We think of this shape as a 4 unit by 5 unit rectangle with two triangles removed, as shown.

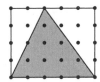

The area of the 4 unit by 5 unit rectangle is $4\times5=20$ square units.

The area of the triangle on the left is half of $4\times2=8$ square units. So, the area of this triangle is $8\div2=4$ square units.

The area of the triangle on the right is half of $3\times4=12$ square units. So, the area of this triangle is $12\div2=6$ square units.

The total area of the shaded region is $20-4-6=$ **10 square units**.

53. We think of this shape as a 5 unit by 5 unit square with two triangles removed, as shown.

The area of a 5-unit unit square is $5\times5=25$ square units.

The area of each triangle is half of a 5 unit by 1 unit rectangle. So, the total area of the two triangles is $5\times1=5$ square units.

The total area of the shape is $25-5=$ **20 square units**.

54. The shape is made of 4 triangles, as shown:

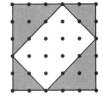

The upper-left and lower-right triangles each have an area that is half of a 3 unit by 3 unit square. So, the total area of these two triangles is $3\times3=9$ square units.

The lower-left and upper-right triangles each have an area that is half of a 2 unit by 2 unit square. So, the total area of these two triangles is $2\times2=4$ square units.

The total area of the shaded region is $9+4=$ **13 square units**.

55. We can split the shape into four triangles and one rectangle, as shown:

The area of the 5 unit by 1 unit rectangle is $5\times1=5$ square units.

The area of each triangle is half of a 2 unit by 2 unit square. We can arrange these four triangles to make two 2 unit by 2 unit squares. The total area of the four triangles is $2\times(2\times2)=2\times4=8$ square units.

The total area of the shaded region is $5+8=$ **13 square units**.

56. We think of this shape as a 5 unit by 5 unit square with four triangles removed, as shown.

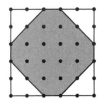

The area of the 5 unit by 5 unit square is $5\times5=25$ square units.

The area of each triangle is half of a 2 unit by 2 unit square. We can arrange these four triangles to make two 2 unit by 2 unit squares. So, the total area of the four triangles is $2\times(2\times2)=2\times4=8$ square units.

The total area of the shaded region is $25-8=$ **17 square units**.

57. We think of this shape as a 4 unit by 5 unit rectangle with three triangles removed, as shown.

The area of the 4 unit by 5 unit rectangle is $4\times5=20$ square units.

The area of the upper-left triangle is half of $2\times5=10$ square units. So, the area of this triangle is $10\div2=5$ square units.

The area of the lower-left triangle is half of $2\times3=6$ square units. So, the area of this triangle is $6\div2=3$ square units.

The area of the lower-right triangle is half of $4\times2=8$ square units. So, the area of this triangle is $8\div2=4$ square units.

The total area of the shaded region is $20-5-3-4=$ **8 square units**.

58. We think of this shape as a 5 unit by 5 unit square with four triangles removed, as shown.

The area of the 5 unit by 5 unit square is $5 \times 5 = 25$ square units.

The area of each triangle is half of a 3 unit by 2 unit rectangle. We can arrange these four triangles to make two 3 unit by 2 unit rectangles. So, the total area of the four triangles is $2 \times (3 \times 2) = 2 \times 6 = 12$ square units.

The total area of the shaded region is $25 - 12 = \textbf{13 square units}$.

59. We can split the shape into twelve pieces: 8 congruent triangles and 4 congruent rectangles, as shown.

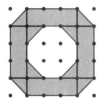

The area of each 3 unit by 1 unit rectangle is $3 \times 1 = 3$ square units. So, the total area of all four rectangles is $3 \times 4 = 12$ square units.

The area of each triangle is half of a 1 unit by 1 unit square. We can arrange these 8 triangles to make four 1 unit by 1 unit squares. So, the total area of the eight triangles is $4 \times (1 \times 1) = 4$ square units.

The total area of the shaded region is $12 + 4 = \textbf{16 square units}$.

60. We think of this shape as a 6 unit by 6 unit square with eight congruent right triangles removed, as shown.

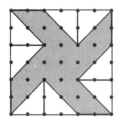

The area of the 6 unit by 6 unit square is $6 \times 6 = 36$ square units.

The area of each triangle is half of a 2 unit by 2 unit square. We can arrange these 8 triangles to make four 2 unit by 2 unit squares. So, the total area of the eight triangles is $4 \times (2 \times 2) = 4 \times 4 = 16$ square units.

The total area of the shaded region is $36 - 16 = \textbf{20 square units}$.

61. We can split the shape into twelve congruent squares, as shown.

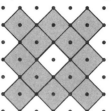

Each square can be split into four congruent triangles, as shown:

The area of each triangle is half of a 1 unit by 1 unit square. We can arrange these 4 triangles to make two 1 unit by 1 unit squares. So, the total area of the four triangles is $2 \times (1 \times 1) = 2 \times 1 = 2$ square units.

So, the area of each congruent square is 2 square units.

The shape is made of 12 squares, each with an area of 2 square units. The total area of the shaded region is $12 \times 2 = \textbf{24 square units}$.

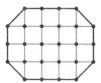

62. The area of the rectangular game board is $4 \times 3 = 12$ square units. The winning move splits this board into two pieces that each have an area of $12 \div 2 = 6$ square units.

We have five possible moves to finish the game:

Of these five moves, only the following line splits the grid into two pieces with equal area:

The shaded region has five whole squares and two half squares, for a total area of $5 + 1 = 6$ square units.

The unshaded region also has an area of $12 - 6 = 6$ square units.

63. The game board has 16 whole squares and 4 half squares, for a total area of $16 + 2 = 18$ square units:

The winning move splits this board into two pieces that each have an area of $18 \div 2 = 9$ square units. We have four possible moves to finish the game:

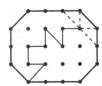

Of these four moves, only the following line splits the grid into two pieces with equal area:

The shaded region has six whole squares and six half squares, for a total area of $6+3=9$ square units.

The unshaded region also has an area of $18-9=9$ square units.

64. The area of the rectangular game board is $4\times5=20$ square units. The winning move splits this board into two pieces that each have an area of $20\div2=10$ square units.

We have three possible moves to finish the game:

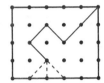

Of these three moves, only the following line splits the grid into two pieces with equal area:

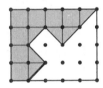

The shaded region has 7 whole squares and six half squares, for a total area of $7+3=10$ square units.

The unshaded region also has an area of $20-10=10$ square units.

65. The shape has fourteen whole squares and four half squares, for a total area of $14+2=16$ square units.

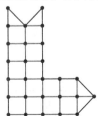

So, we look for the fewest possible moves to split the shape into two pieces that each have an area of $16\div2=8$ square units.

We split the shape into two pieces of equal area with **two** moves as shown:

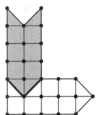

The shaded region has six whole squares and four half squares, for a total area of $6+2=8$ square units.

The unshaded region also has an area of $16-8=8$ square units.

We cannot split the shape into two pieces of equal area with fewer than **2** moves.

66. The shape has twenty-one whole squares and two half squares, for a total area of $21+1=22$ square units.

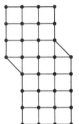

So, we look for the fewest possible moves to split the shape into two pieces that each have an area of $22\div2=11$ square units.

We split the shape into two pieces of equal area with **three** moves as shown:

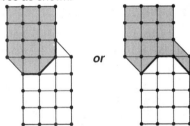

In the left diagram above, the shaded region has ten whole squares and two half squares, for a total area of $10+1=11$ square units.

In the right diagram above, the shaded region has nine whole squares and four half squares, for a total area of $9+2=11$ square units.

In both diagrams, each unshaded region also has an area of $22-11=11$ square units.

We cannot split the shape into two pieces of equal area with fewer than **3** moves.

67. The area of the shape is 22 square units.

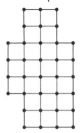

So, we look for the fewest possible moves to split the shape into two pieces that each have an area of $22 \div 2 = 11$ square units.

We split the shape into two pieces of equal area with **three** moves as shown:

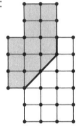

The shaded region has ten whole squares and two half squares, for a total area of $10 + 1 = 11$ square units.

The unshaded region also has an area of $22 - 11 = 11$ square units.

We cannot split the shape into two pieces of equal area with fewer than **3** moves.

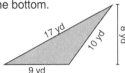

68. The base length of this triangle is 8 feet. The height of this triangle is 10 feet. So, the area of the triangle is $8 \times 10 \div 2 = 80 \div 2 = $ **40 sq ft**.

69. The base length of this triangle is 15 inches. The height of this triangle is 10 inches. So, the area of the triangle is $15 \times 10 \div 2 = 150 \div 2 = $ **75 sq in**.

70. The base length of this triangle is 21 cm. The height of this triangle is 12 cm. So, the area of the triangle is $21 \times 12 \div 2 = 252 \div 2 = $ **126 sq cm**.

71. The height of the triangle is measured from the side whose length is 9 yards, so we turn the triangle so that this side is on the bottom.

The base length of this triangle is 9 yards. The height of this triangle is 8 yards. So, the area of the triangle is $9 \times 8 \div 2 = 72 \div 2 = $ **36 sq yd**.

— *or* —

We do not need to turn the triangle to calculate its area! Since the height of the triangle is measured from the side whose length is 9 yards, the base length is 9 yards. The height is 8 yards. So, the area of the triangle is $9 \times 8 \div 2 = 72 \div 2 = $ **36 sq yd**.

72. The area of a triangle is half the product of its base length and height:

$$\text{Area} = \text{base} \times \text{height} \div 2$$
$$(A = b \times h \div 2)$$

Since the area of the triangle is 45 sq ft and the base length is 15 ft, we have

$$45 = 15 \times h \div 2$$

Since *half* of $15 \times h$ is 45 sq ft, we know that $15 \times h$ is $45 \times 2 = 90$ sq ft:

$$90 = 15 \times h$$

Since $90 = 15 \times \boxed{6}$, we have $h = 6$. So, a triangle with an area of 45 sq ft and base length 15 ft has height **6 ft**.

— *or* —

The area of a triangle is half the area of a rectangle that has the same base length and height as the triangle. So, a rectangle with the same base length and height as the given triangle will have twice the area of the triangle: $45 \times 2 = 90$ sq ft.

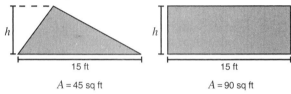

A rectangle with an area of 90 sq ft and base length 15 ft has height $90 \div 15$ feet. Since $90 = 15 \times \boxed{6}$, we have $90 \div 15 = 6$. So, the height of the rectangle is 6 feet.

The triangle has the same height. So, a triangle with an area of 45 sq ft and base length 15 ft has height **6 ft**.

We check our answer:
The area of a triangle with base length 15 ft and height 6 ft is $15 \times 6 \div 2 = 90 \div 2 = 45$ sq ft.

73. We use a ruler to measure the base and height of the triangle, using the long side of the triangle as the base:

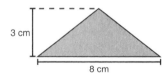

Then, we calculate the area of the triangle:

$$8 \times 3 \div 2 = 24 \div 2 = \textbf{12 sq cm}.$$

74. The height of each triangle is 8 inches. The sum of the base lengths of three congruent triangles is 21 inches. So, the base length of one triangle is $21 \div 3 = 7$ inches:

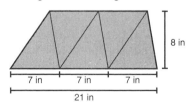

The area of one triangle with base length 7 inches and height 8 inches is $7 \times 8 \div 2 = 56 \div 2 = 28$ sq in.

Therefore, the area of the quadrilateral formed by the 5 congruent triangles is $5 \times 28 = $ **140 sq in**.

75. We use the bottom side of each triangle as the base, and we use h to represent the height of each triangle from that base:

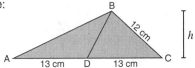

Triangles ABD and BCD each have base length 13 cm and height h. Two triangles with the same height and base length have the same area. So, the area of triangle ABD is equal to the area of triangle BCD. This means that the area of each smaller triangle is equal to half the area of triangle ABC.

The area of triangle ABC is 104 sq cm. So, the area of triangle ABD is $104 \div 2 = $ **52 sq cm**.

AREA
Fragment Puzzles   96-97

76. The area of a 3 unit by 4 unit rectangle is $3 \times 4 = 12$ square units. So, the three pieces used to make the rectangle must have a combined area of 12 square units.

In the four pieces shown, there are a total of 14 whole squares and 8 half squares, for a total area of $14 + 4 = 18$ square units. That is 6 more square units than we need!

If we remove the piece that has an area of 6 square units, then we are left with three pieces whose total area is $18 - 6 = 12$ square units.

Only the piece on the right (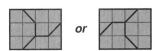) has an area of 6 square units (5 whole squares + 2 half squares).

The other three pieces can be arranged to create a 3 unit by 4 unit rectangle, as shown below:

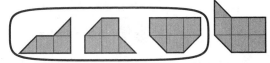

We circle the three pieces used:

77. The area of the target shape is 6 square units (5 whole squares + 2 half squares).

In the four pieces shown, there are 3 whole squares and 8 half squares, for a total area of $3 + 4 = 7$ square units. That is 1 square unit more than we need!

Only the piece on the left ( ) has an area of 1 square unit (2 half squares). The other three pieces can be arranged to make the target shape, as shown below:

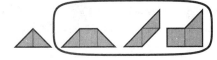

We circle the three pieces used:

78. The area of the target shape is 11 square units (10 whole squares + 2 half squares).

In the four pieces shown, there are 8 whole squares and 12 half squares, for a total area of $8 + 6 = 14$ square units. That is 3 square units more than we need!

Only the third piece from the left (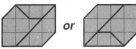) has an area of 3 square units (2 whole squares + 2 half squares). The other three pieces can be arranged to make the target shape, as shown below:

We circle the three pieces used:

79. The area of the target shape is 12 square units (10 whole squares + 4 half squares).

In the four pieces shown, there are 13 whole squares and 10 half squares, for a total area of 18 square units. That is 6 square units more than we need!

Only the third piece from the left ( ) has an area of 6 square units (5 whole squares + 2 half squares). The other three pieces can be arranged to create the target shape, as shown below:

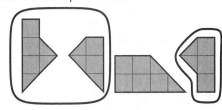

We circle the three pieces used:

80. The area of the target shape is 7 square units (5 whole squares + 4 half squares).

In the four pieces shown, there are 5 whole squares and 8 half squares, for a total area of 9 square units. That is 2 square units more than we need!

Only the second piece from the left (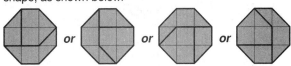) has an area of 2 square units (1 whole square + 2 half squares). The other three pieces can be arranged to create the target shape, as shown below:

We circle the three pieces used:

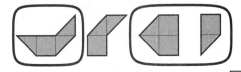

81. A square with an area of 1 square meter has sides that are 1 meter long. Since there are 100 centimeters in a meter, the sides of the square are 100 centimeters long.

So, there are $100 \times 100 = $ **10,000** square centimeters in 1 square meter.

82. A square with an area of 1 square mile has sides that are 1 mile long. There are 5,280 feet in 1 mile, so there are $5,280 \times 5,280$ square feet in 1 square mile.

Since we are asked to circle the number *closest* to the number of square feet in a square mile, we estimate the product $5,280 \times 5,280 \approx 5,000 \times 5,000 = 25,000,000$.

Our estimate suggests that the number of square feet in a square mile ($5,280 \times 5,280$) is closer to **25,000,000** than to any of the other four numbers.

5,000   25,000   5,000,000   (25,000,000)   250,000,000

In fact, the number of square feet in a square mile is $5,280 \times 5,280 = 27,878,400$.

83. A square with an area of 1 square yard has sides that are 1 yard long. There are 3 feet in 1 yard and 12 inches in 1 foot, so there are $3 \times 12 = 36$ inches in one yard.

Since there are 36 inches in one yard, the sides of the square are 36 inches long. So, there are 36×36 square inches in one square yard.

We can underestimate $36 \times 36 \approx 30 \times 30 = 900$.
We can also overestimate $36 \times 36 \approx 40 \times 40 = 1,600$.

So, the number of square inches in one square yard (36×36) is between **900 and 1,600**.

0 and 100   100 and 400   400 and 900   (900 and 1,600)

In fact, the number of square inches in one square yard is $36 \times 36 = 1,296$.

84. A square with an area of 1 square yard has sides that are 1 yard long. Since there are 3 feet in one yard, there are $3 \times 3 = 9$ square feet in one square yard.

So, there are $4 \times 9 = $ **36 square feet** in four square yards.

85. To find the area of the rectangular sidewalk, we can multiply its width by its length: 1 yard × 1 mile.

Since we want the final answer in square *feet*, we first convert each given measurement to feet:
The width of the sidewalk is 1 yard = 3 feet.
The length of the sidewalk is 1 mile = 5,280 feet.

So, the area of the sidewalk is
$3 \times 5,280 = $ **15,840** square feet.

86. Since each square foot has 12-inch sides and each square tile has 4-inch sides, $12 \div 4 = 3$ tiles can be laid against each side of a square foot, as shown:

So, Rosencrantz and Guildenstern can use $3 \times 3 = 9$ tiles to cover each square foot.

An 8 foot by 10 foot floor contains $8 \times 10 = 80$ square feet. So, it takes $9 \times 80 = $ **720** four-inch square tiles to cover 80 square feet.

87. One square yard equals $3 \times 3 = 9$ square feet. Since one ounce of paint covers 3 square feet, Captain Kraken will need $9 \div 3 = 3$ ounces of paint to cover each square yard. So, to cover 15 square yards, he will need $3 \times 15 = $ **45** ounces of paint.

— *or* —

One square yard equals $3 \times 3 = 9$ square feet. So, 15 square yards equal $9 \times 15 = 135$ square feet. Each ounce of paint covers 3 square feet. So, to cover 135 square feet, Captain Kraken will need $135 \div 3 = $ **45** ounces of paint.

88. The width of the table is 3 feet, or $3 \times 12 = 36$ inches. The short side of an index card is 3 inches long. So, we can place $36 \div 3 = 12$ cards along this side, as shown:

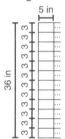

The length of the table is 5 feet, or $5 \times 12 = 60$ inches. The long side of an index card is 5 inches long. So, we can place $60 \div 5 = 12$ cards along its length, as shown:

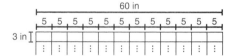

So, we can cover the table with $12 \times 12 = $ **144 cards**, as shown:

89. One square yard equals $3 \times 3 = 9$ square feet of fabric.

Since Ms. Levans can only purchase whole numbers of square *yards*, the number of square feet of fabric she buys must be a multiple of 9.

Ms. Levans needs 50 square feet of fabric.
50 is between $9 \times 5 = 45$ and $9 \times 6 = 54$.

So, to purchase at least 50 square feet of fabric, Ms. Levans must purchase 54 square feet of fabric.
This is $54 \div 9 = $ **6 square yards** of fabric.

90. Since we want to know the number of inches in the width of Cammie's paper, we write the area in square *inches* (instead of square feet). There are $12 \times 12 = 144$ square inches in 1 square foot.

We can find the width of Cammie's paper by dividing its area by its height: $144 \div 8 = 18$.

So, Cammie's paper is **18 inches** wide.

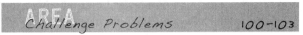

91. The shape is made by attaching two congruent right triangles. The area of each triangle is half the area of a 40 foot by 30 foot rectangle. So, the combined area of the two triangles is $40 \times 30 = $ **1,200 sq ft**.

92. The area of the square is $6 \times 6 = 36$ sq km.
The triangle has the same area as the square.

Since the area of the triangle is 36 sq km, and the base length is 6 km, we have:

$$36 = 6 \times h \div 2$$

Since *half* of $6 \times h$ is 36 sq km, we know that $6 \times h$ is $36 \times 2 = 72$ sq km:

$$72 = 6 \times h$$

Since $72 = 6 \times \boxed{12}$, we have $h = 12$. So, the height of a triangle that has an area of 36 sq km and a base length of 6 km is **12 km**.

— *or* —

The area of a triangle is half the area of a rectangle that has the same base length and height as the triangle. So, a rectangle with the same base length and height as the given triangle will have twice the area of the triangle: $36 \times 2 = 72$ sq km.

The height of a rectangle whose area is 72 sq km and whose base length is 6 km is $72 \div 6 = $ **12 kilometers**. This is the same as the height of the triangle. We check our answer:
The area of a triangle with base length 6 km and height 12 km is $6 \times 12 \div 2 = 72 \div 2 = 36$ sq km.

93. Since the side length of EFGH is twice the side length ABCD, we can arrange 4 copies of square ABCD to make a square the size of EFGH, as shown:

So, the area of EFGH is four times the area of ABCD: $7 \times 4 = $ **28 sq m**.

94. The area of a 10 foot by 18 foot rug is $10 \times 18 = 180$ sq ft.

One square yard equals $3 \times 3 = 9$ square feet. So, in 180 square feet, there are $180 \div 9 = $ **20 square yards**.

95. Square PQRS is divided into 8 congruent triangles. We can pair these triangles to make four small, congruent squares, as shown:

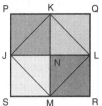

Since the combined area of the 4 small squares is the same as the area of PQRS (28 sq ft), the area of each small square is $28 \div 4 = 7$ sq ft.

So, the area of square PKNJ is **7 sq ft**.

96. Square PQRS is divided into 8 congruent triangles, and square JKLM is made of four of these congruent triangles. From 8 congruent triangles, we can make 2 groups of four triangles. For example,

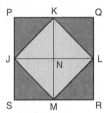

Since the combined area of the 8 triangles is 28 sq ft, each group of four triangles has a total area of $28 \div 2 = 14$ sq ft.

So, the area of square JKLM is **14 sq ft**.

— *or* —

The area inside of square JKLM is the same as the area that is inside of PQRS but outside of JKLM:

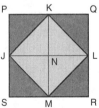

Since these two areas are equal, the area of JKLM is half the area of PQRS: $28 \div 2 = $ **14 sq ft**.

97. The square is made by attaching the 4-inch side of one triangle to the 9-inch side of another triangle.

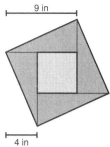

So, the side length of the small square is $9-4=5$ inches.

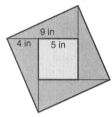

The area of a square with side length 5 inches is $5\times5=$ **25 sq in**.

98. The area of the large square is equal to the sum of the areas of the four congruent triangles plus the area of the small square.

The area of each congruent right triangle is $9\times4\div2=36\div2=18$ sq in. So, the total area of the four congruent triangles is $4\times18=72$ sq in.

In the previous problem, we calculated the area of the small square: 25 sq in.

So, the area of the large square is $72+25=$ **97 sq in**.

99. Turning the triangle on its side does not change its area. So, when we compute the area of the triangle using the 4-inch side as its base, we get the same area as when we compute the area using the 20-inch side as its base.

$$4\times15\div2=20\times h\div2$$

We see that half of 4×15 equals half of $20\times h$. This means that 4×15 and $20\times h$ are equal.

$$4\times15=20\times h$$

We know $4\times15=60$, so we have $60=20\times h$. Then, since $60=20\times\boxed{3}$, we have $h=3$. So, the height of the triangle when measured from the 20-inch side as its base is **3 inches**.

Note that multiplying the base length of a triangle by its height will always give the same product, no matter which side we use as the base of the triangle.

100. The area of an 8-foot square is $8\times8=64$ sq ft. We calculate the area of each triangle:

 $2\times8\div2=16\div2=8$ sq ft

 $3\times8\div2=24\div2=12$ sq ft

$4\times8\div2=32\div2=16$ sq ft

 $5\times8\div2=40\div2=20$ sq ft

 $6\times8\div2=48\div2=24$ sq ft

The total area of all five triangles is $8+12+16+20+24=80$ sq ft.

That is $80-64=16$ square units more than we need!

So, we remove the triangle with an area of 16 sq ft (the triangle with base 8 ft and height 4 ft). The area of the remaining 4 triangles is $8+12+20+24=64$ sq ft.

We can arrange those 4 triangles into an 8-foot square as shown:

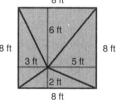

We circle the four triangles used above:

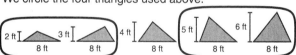

101. The area of a triangle is half the area of a rectangle that has the same base length and height as the triangle.

So, for a triangle and a rectangle with the same base length to have the same area, the triangle must be twice as tall as the rectangle. For example:

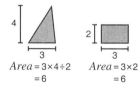

$Area=3\times4\div2$ $Area=3\times2$
  $=6$ $=6$

Since the triangle in the problem has the same base length and area as the square, it is twice as tall as the square. So, we can label the height of the triangle $2\times s$:

$2\times s$

s

s s

$Area=2\times s\times s\div2$ $Area=s\times s$
   $=s\times s$

So, the height of the pentagon is the same as the height of three squares with side length s:

Since the height of the pentagon is 24 m, we have $s = 24 \div 3 = 8$ m.

The side length of the square is **8 meters**.

102. We calculate the area of the square hole by subtracting the areas of the 4 congruent right triangles from the area of the 10-meter square they create.

The area of a 10-meter square is $10 \times 10 = 100$ sq m. This is the shaded area below:

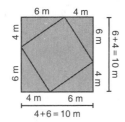

The area of each congruent triangle is $4 \times 6 \div 2 = 24 \div 2 = 12$ sq m. The total area of the four triangles is therefore $4 \times 12 = 48$ sq m. Subtracting the areas of the four triangles, we are left with the shaded area below:

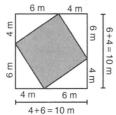

This is the area of the square hole: $100 - 48 = $ **52 sq m**.

103. The two triangles we make will have the same height. They also must have the same area. The area of a triangle is:

$$\text{Area} = \text{base} \times \text{height} \div 2.$$

So, two triangles with the same height and the same area must have the same base length.
Since the length of the bottom side of the original triangle is 6 units, the two smaller triangles must have base length $6 \div 2 = 3$ units.

We split the triangle into two smaller triangles with equal area by drawing the following line:

104. The area of the original square is 16 square inches, so the side length of the original square is 4 inches.

So, we can place the original square on a 4 inch by 4 inch dot grid, as shown:

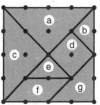

We split this grid into 1-inch squares:

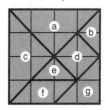

a. This tan has two squares and four half squares for a total area of **4 sq in**.

b. This tan has two half squares for a total area of **1 sq in**.

c. This tan has two whole squares and four half squares for a total area of **4 sq in**.

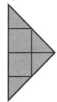

d. This tan has four half squares for a total area of **2 sq in**.

e. This tan has two half squares for a total area of **1 sq in**.

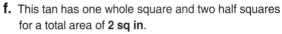

f. This tan has one whole square and two half squares for a total area of **2 sq in**.

g. This tan has one whole square and two half squares for a total area of **2 sq in**.

We check the total area of these seven pieces:
$$4 + 1 + 4 + 2 + 1 + 2 + 2 = 16 \text{ sq in}.$$

For additional books, printables, and more, visit
www.BeastAcademy.com